# Introduction to Laser Science and Engineering

# Introduction to Laser Science and Engineering

Travis S. Taylor

## CRC Press
Taylor & Francis Group
Boca Raton London New York

CRC Press is an imprint of the
Taylor & Francis Group, an **Informa** business

CRC Press
Taylor & Francis Group
6000 Broken Sound Parkway NW, Suite 300
Boca Raton, FL 33487-2742

**Library of Congress Cataloging in Publication Data**

Names: Taylor, Travis S., author.
Title: Introduction to laser science and engineering / Travis S. Taylor.
Description: First edition. | Boca Raton, FL: CRC Press/Taylor & Francis Group,
    2020. | Includes bibliographical references and index.
Identifiers: LCCN 2019014815 | ISBN 9781138036390 (hardback: acid-free paper) |
    ISBN 9781315178561 (ebook)
Subjects: LCSH: Lasers.
Classification: LCC TA1675 .T39 2020 | DDC 621.36/6—dc23
LC record available at https://lccn.loc.gov/2019014815

**Visit the Taylor & Francis Web site at
http://www.taylorandfrancis.com**

**and the CRC Press Web site at
http://www.crcpress.com**

eResource material is available for this title at https://www.crcpress.com/9781138036390

# Contents

# List of Figures

# List of Tables

# *Preface*

This book was written with the hope of creating a source for an undergraduate, technical professional from another field, or learned enthusiast that will give them exposure to the field of laser science and engineering and some idea as to the full breadth of knowledge the topic encompasses. The idea behind this text was to create a different approach from the standard laser textbooks, which are usually at the graduate level written for specialists in the fields of optics, physics, or electrical engineering.

I began working with lasers at the age of 17 for what was then called the U.S. Army Directed Energy Directorate in the summer of 1986 and it was clear to me then that it was going to take decades to learn all I needed to learn about lasers to be an effective researcher with them. Back then, the best textbook was an elder mentor in the field who had learned lessons from hard knocks, a few mistakes, and painstaking hours, days, weeks, months, and even years to become an expert in the field of laser science and engineering. At the time, I spent every moment I could, listening to their stories, advice, impromptu lectures, and most importantly their safety hints and habits.

The first time I learned to align a laser using another lower-powered laser and many mirrors, irises, pinholes, and index cards took me several days to get it right even after being shown how to do it. I have never seen such processes mentioned in textbooks and seldom in lab classes at the college graduate level. But on the occasion where an elder laser scientist or engineer would show me a trick like looking for interference rings on the face of the alignment laser, using first-surface reflections off of prism faces, using thermal printer paper to find an infrared beam, or using a bleached cloth to find an ultraviolet beam, I was truly learning what it meant to become a laser scientist or an engineer.

This book hopefully is written in a way to relay some of these lessons learned as well as the general approach of looking at lasers from a slightly more pragmatic perspective. Lasers are all around us; every day from the grocery store checkout counter to the DVD drives and players in our homes, to games, to hospitals and medical clinics, to optical internet and communications, and to weapons. It is very likely that lasers will be part of humanity's technological culture for quite a long time and the need for each new generation to study, experiment with, and apply them will be as well.

Hopefully, this book will get you started with a holistic view of what lasers are and it will also supply some practical knowledge, application, and safety information that I have learned over the years. While there are more complete and detailed books on lasers out there, they were not designed as an introduction—a true first exposure to the what and how of lasers. Work through this book and then begin to apply the knowledge in the lab and the real world. Then start in on those more detailed texts knowing that you have built a good basic understanding. So, good luck, and let's get started with an *Introduction to Laser Science and Engineering.*

# *Introduction*

Since Charles Townes and Arthur Schawlow conceived of the basic ideas for the laser in 1958, Gordon Gould coined the acronym LASER, and Theodore Maiman demonstrated the first actual laser in 1960; the laser has become a staple in our modern technological culture. The laser is found in almost every home in DVD players, video game consoles, security systems, and in neighborhoods where modern-day fiber optical cables bring Internet data of laser beams. The laser is most likely here to stay for a long time.

In many cases, and for most people, the laser is one of those magical widgets that we trust like a light switch. We flip a switch and we expect lights to come on or go off. Most of the general public has no real understanding of how the power is generated and delivered to our homes via electromagnetic signal transmission and then transformed into a useful form before we can use it, but we do expect the lights to come on when we flip the switch. Lasers have become the same for most of us. When presenters are giving presentations, talks, or lectures they expect that a laser pointer will place a dot on the screen at the press of a button, and little, if no, thought is given to the fundamental quantum physics processes that are taking place in the palm of their hand.

But somebody had to invent that laser. Somebody had to understand those fundamental quantum physics processes taking place. Somebody had to learn how to construct the device that could take electrical power from a battery and convert it into a coherent beam of visible light. That somebody would have been a laser scientist or an engineer. This book is written for those who desire to be that somebody someday. The book is written to be a first exposure to the modern physical knowledge required to understand basic laser science and some useful and practical engineering and applications of them. This book is an *Introduction to Laser Science and Engineering* that should be a good starting point for the undergraduate, the technical professional from another field, or the talented enthusiast.

# *Author*

**Travis S. Taylor** ("Doc" Taylor to his friends) earned his soubriquet the hard way: He has a doctorate in Optical Science and Engineering, a doctorate in Aerospace Systems Engineering, a master's degree in Physics, and a master's degree in Aerospace Engineering, all from the University of Alabama in Huntsville. Added to this are a master's degree in Astronomy from the University of Western Sydney (Australia) and a bachelor's degree in Electrical Engineering from Auburn University (Alabama). Dr. Taylor has worked on various programs for the Department of Defense and NASA for the past two decades. He is currently working on several advanced propulsion concepts, very large space telescopes, space-based beamed energy systems, next-generation space launch concepts, directed energy weapons, nanosatellites, and low-cost launch vehicle concepts for the U.S. Army Space and Missile Defense Command. Dr. Taylor was one of the principal investigators of the U.S. Army's 100 Joule Tunable Laser program, narrow linewidth dye laser research program, multi-wavelength oscillator-amplifier experiment, ruggedized flashlamp-pumped dye laser experiment, and the solid-state laser testbed experiment. He is currently researching high brightness quantum entanglement sources for space communications.

In his copious spare time, Doc Travis is also a black belt martial artist, a private pilot, a SCUBA diver, and has raced mountain bikes. He has also competed in triathlons, is a marathon runner, a CrossFitter, and has been the lead singer and rhythm guitarist of several hard rock bands. He has written about two dozen science fiction novels, three textbooks (including this one), and over a dozen refereed technical papers. Dr. Taylor has appeared and starred in several television programs including the History Channel's *The Universe, Life After People, Ancient Aliens, The Curse of Oak Island*, and *Rise of the Superbombs*, National Geographic Channel's hit shows *Rocket City Rednecks* and *When Aliens Attack*, Science Channel's *NASA's Unexplained Files*, and The Weather Channel's *3 Scientists Walk Into a Bar*. He currently lives with his wife and two children in north Alabama just outside of Huntsville in view of the Saturn V rocket that is erected at the U.S. Space and Rocket Center.

# 1

## What Is Light?

This book is entitled *Introduction to Laser Physics and Engineering*, so one would suspect that like all the other introductory laser books out there we would start with talking about what lasers are and jump right into the concept of stimulated emission. Looking through most texts on the subject to date, more than most mention stimulated and spontaneous emission on the very first page. We've mentioned it here so that it puts this book at least on par for that course. That said, it is the author's viewpoint that jumping right into the discussion of stimulated emission and spontaneous emission without taking a few steps back and getting our understanding of the more basic concepts involving lasers straight in our minds we might actually miss how truly profound lasers are and how much we can learn from them. In fact, many texts on the subject start with semi-classical physical nomenclature that is neither illuminating nor completely accurate which can become confusing for the entry-level laser scientist or engineer. Hopefully, this book will offer a slightly different approach that will in the end (pun intended) shed more light on the topic, be enlightening (another intended pun), and enable us to obtain a laser tight focus (that one just happened, honestly) on such an intricate, detailed, complicated, and highly exciting subject.

LASER is actually an acronym just like NASA, radar, lidar, and ASAP. We all know that NASA is the National Aeronautics and Space Administration, radar is a U.S. Navy acronym for Radio Detection (or Direction) and Ranging, and ASAP means As Soon As Possible. Lidar is a bit more obfuscated in that it is an attempt at creating a word like "radar" but for lasers and/or light. Lidar is often described as Light Detection and Ranging, while sometimes it is Laser Imaging, Detection, and Ranging. The first one is the most widely used because not all lidar systems perform imaging.

LASER actually stands for Light Amplification by the Stimulated Emission of Radiation. So, if we wanted to be absolutely nitpicky and exactly correct with our acronym instead of "laser," it would be "labtseor" but the latter doesn't quite roll off the tongue as well as the former. Hence, laser. Also, note that "laser" has become such a common acronym that it has become grammatically acceptable to write it in lowercase just like "radar."

Looking at the acronym itself is the simplest method to determine where to start in our study of the laser. The very first word in the acronym is "light." So, right off the bat we can ask ourselves one simple question and if we can't answer it succinctly then we have some work to do before we can go any further. So, let's ask that question (and hopefully answer it) ASAP.

What is light?

## 1.1 The Classical Description of Light

### 1.1.1 500 to 0 B.C.E.

For centuries the nature of light has been debated, discussed, and experimented with. There is no singular truly "classical description" of light because there are several. In the 5th century B.C.E., the pre-Socratic Greek philosopher Empedocles (see Figure 1.1) of Acragas (a city in Sicily) developed what became known as the *cosmogenic theory* of the universe where he proposed that the universe was made from four classical elements: fire, air, water, and earth. He believed that the goddess Aphrodite had made the human eye from these elements and that she lit the fire within the eye, which in turn, shone outward illuminating objects and therefore leading to sight. Empedocles was clever enough to realize if this were true then humans would be able to see at night, so, he postulated that there was somehow an interaction of "rays" from the eyes and "rays" from the sun or other sources.

Empedocles.

**FIGURE 1.1**
Empedocles of Acragas theorized that light was given to man by the goddess Aphrodite, who had made the human eye from these elements and that she lit the fire within the eye, which, in turn, shone outward illuminating objects and therefore leading to sight.

The ancient Hindu text Vishnu Purana that was written along the same time frame, give or take a couple 100 years based on the particular estimate or history book, describes the "seven rays of the sun." As far back as ancient civilization, light was being referred to in forms of rays.

The Greek philosopher Democritus of Abdera (see Figure 1.2), around 400 B.C.E., argued that objects generated "simulacra" of themselves or images that were cast-off of them and into space. When these simulacra were impressed upon the surrounding air, in turn, the impression was then fed to the eye. Most scientist and engineers with a modern education in light will almost laugh at such a description of light. However, we will see later in this chapter that this idea is actually not completely laughable.

Specifically, the ancient Greek philosophers Socrates, Plato, and Aristotle (see Figure 1.3) developed the basic foundations for most of the modern sciences, mathematics, and philosophy between 469 and 322 B.C.E. But these philosophers disagreed with Democritus and still believed that sight was transmitted from the eye rather than received there. They eventually came to what is sometimes referred to as a "classical compromise" where it was argued that it took both the external simulacra or emitted image that then interacted with the internal "inner fire" of the eye that caused the act of seeing. What is interesting here is how close to modern quantum theory this is, which will be discussed later in this chapter. Again, an ancient theory that on the surface seems laughable actually isn't too far from the mark. Suffice it to say that the modern accepted "classical" theory that is taught at the undergraduate and even graduate level is that light is external and incident on the eye. Again, that is the modern "classical" interpretation.

**FIGURE 1.2**
The Greek philosopher Democritus of Abdera, around 400 B.C.E., argued that objects generated "simulacra" of themselves or images that were cast-off of them and into space.

**FIGURE 1.3**
The ancient Greek philosophers Socrates, Plato, and Aristotle developed the basic foundations for most of the modern sciences, mathematics, and philosophy between 469 and 322 B.C.E. But these philosophers disagreed with Democritus and still believed that sight was transmitted from the eye rather than received there.

Figure 1.4 shows Euclid who then (around 300 B.C.E.) built on the ancient Greek philosophers' foundation and authored the five works *Elements, Data, Catoptrics, Phaenomena, and Optics*. It is within these works that Euclid describes for the first time concepts such as geometry systems and coordinate frames, mathematical theory of mirrors, spherical astronomy, as well

**FIGURE 1.4**
Euclid who then (around 300 B.C.E.) built on the ancient Greek philosophers' foundation, authored the five works *Elements, Data, Catoptrics, Phaenomena*, and *Optics*.

as reflection and diffusion and their relationships with vision. Without Euclidian geometry, we would neither be able to discuss the path of beams of light along an "optical axis" nor be able to discuss the interactions between beams at various points in space. Euclid gave us a graphical geometrical way to envision the universe around us and the things within it.

In the ancient era, more and more thinkers such as Euclid and then Titus Lucretius Carus, a Roman poet and philosopher, who authored the poem De rerum natura, as shown in Figure 1.5, which is translated as "On the Nature of Things" or "On the Nature of the Universe" began to discuss the idea of a sight and why humanity could see. It was within this manuscript where Titus used poetic language to explain the ideas of Epicureanism, which includes atomism and psychology. Titus was the first writer to introduce Roman readers to such concepts as light being tiny atoms of energy and both he and Euclid questioned the idea that sight didn't emit from the eyes because as soon as one opens his/her eyes he/she can see the stars. How could one see the stars from so far away if the light had to leave the eye and travel to the star? This could only be possible if light were infinitely fast. Isn't it?

In 214–212 B.C.E. Archimedes was said to have built a "heat ray" to burn ships at the Siege of Syracuse (Figure 1.6). In many historical and poetic pieces of the era, it was said that Archimedes destroyed the Roman ships with fire. Various descriptions of his device describe it as a "burning glass" with "many moving parts." It was only until the modern era where various

**FIGURE 1.5**
Titus Lucretius Carus, a Roman poet and philosopher, authored the poem *De rerum natura* translated as "On the Nature of Things" or "On the Nature of the Universe" first proposed light as tiny "atoms."

**FIGURE 1.6**
In 214–212 B.C.E., Archimedes was said to have built a "heat ray" to burn ships at the Siege of Syracuse.

television personalities made attempts to discredit the Archimedes heat ray; however, ancient historians clearly described it as successful in setting fire to the Roman fleet. The ancient historian Pappus described the Siege of Syracuse as follows:

> When Marcellus had placed the ships a bow shot off, the old man (Archimedes) constructed a sort of hexagonal mirror. He placed at proper distances from the mirror other smaller mirrors of the same kind, which were moved by means of their hinges and certain plates of metal. He placed it amid the rays of the sun at noon, both in summer and winter. The rays being reflected by this, a frightful fiery kindling was excited on the ships, and it reduced them to ashes, from the distance of a bow shot. Thus the old man baffled Marcellus, by means of his inventions.

It should also be noted that the U.S. Army has an internal-directed energy course where the "heat ray" is taught as the first successfully directed energy weapon (your author has both taken and taught the said course). Just because television special effects teams can't reproduce it doesn't mean it was a myth. In fact, reading the historical records (just the paragraph above) when compared to the methods attempted by many of today shows us that the modern attempts were all wrong. Most of them use flat mirrors and then point those mirrors all at the same location on the mockup boats. The paragraph above tells us that there were multiple moving parts with hinges and plates of metal. It is likely that Archimedes used these hinges and plates of metal to warp the mirrors into adjustable parabolic mirrors. And as history has recorded, he used them to focus the sunlight onto the Roman fleet and set them ablaze. While this heat ray wasn't a laser, it was indeed the first use of light rays as a weapon and it shows that Archimedes brought forth a new understanding of the applications and manipulation of light. And it was the first successful use of directed-energy weapons over 2000 years ago.

It is interesting to note here that in many textbooks to date we will find statements about how wrong the Greek philosophers were about the nature of light and seeing, but how important their early works were in leading us to our modern understanding. In *The Physics of Invisibility: A Story of Light and Deception* (M. Beech, 2012) the following is stated about these philosophers and their view on light:

> This dual – two kinds of light, internal and external – approach to seeing dominated intellectual thought for the best part of a thousand years, and it was not until the early Middle Ages that alternatives were actively pursued.
> Although the classical Greek notion of what light is turned out to be quite wrong, the concept was not damaging to the search for the correct interpretation and development of ideas concerning the workings of light and its interactions with matter.

In as late as 2012, the author of this book claims that the ancient philosophers were "quite wrong" in their description of light. We will see as our research of light's history progresses that this is not actually the case and that most text and modern undergraduate explanations of light are very simplistic and indeed incomplete to a fault. But that will come later.

### 1.1.2 0–200 A.D.

At some point around 100 A.D., Hero of Alexandria (see Figure 1.7) was the first to point out that light always travels in the shortest optical path distance. He is also credited with using this information to explain the law

**FIGURE 1.7**
Around 100 A.D. Hero of Alexandria developed the concept that light travels the shortest optical path distance between two points.

**FIGURE 1.8**
Between 100 and 170 A.D. Claudius Ptolemy, another Greek thinker, created Optics, where he discusses the properties of light.

of reflection through a mathematical proof rather than purely experimental data. He is known to have written many volumes on various science and engineering topics. His work *Catoptrica* was the most pertinent as it covered the subjects of the progression of light, reflection, and the use of mirrors.

Between 100 and 170 A.D. Claudius Ptolemy (see Figure 1.8), another Greek thinker, created *Optics*, where he discusses the properties of light. In this work, he describes reflection and refraction as well as color. The work is considered to be significant historically mostly in that it was a major influence to future works in the area. It specifically influenced Ibn al-Haytham's work in the 11th century. Ptolemy's book was the first to actually have a table showing refraction angles as light passes from air to water. He also offered explanations of other light-based phenomena such as image size, shape, and movement as well as the idea of binocular vision (which humans and most animals have).

This era between the ancient philosophers and Hero was the beginning of applying geometry and mathematics to the scientific description for light rather than simply metaphysics and philosophy. However, many of the mathematical tools required to describe the physics of light had yet to be discovered and indeed it would be some time before it would be.

### 1.1.3 801–873 A.D.

Abu Yusuf Ya'qub ibn 'Ishaq as-Sabbah al-Kindi (see Figure 1.9) was an Arab Muslim philosopher and multitalented polymath known as the "father of Arab philosophy." His approach was to take the works of the Greek and Hellenistic philosophies and meld them into a philosophy he introduced to the Muslim world. As far as light and optics are concerned, he examined both

**FIGURE 1.9**
Abu Yusuf Ya'qub ibn 'Ishaq as-Sabbah al-Kindi was an Arab Muslim philosopher and a multitalented polymath known as the "father of Arab philosophy."

Aristotle's and Euclid's theories on seeing and decided that Euclid's must be
more correct. His argument for this was that Aristotle had suggested that
seeing was caused by the transmission of a "simulacra" of the object being
seen. So were this theory to be correct, according to al-Kindi, then seeing an
object should enable seeing the entirety of an object no matter the angle. For
example, seeing a coin on edge it appears as a line, but seeing the coin's face it
appears as a circle. al-Kindi believed that if Aristotle were correct one would
see the entire coin no matter the viewpoint.

On the other hand, Euclid had described a geometrical theory based on
rays and angles and seeing would be angle specific to the viewpoint of
an object. Therefore, al-Kindi concluded that Euclidean optics was more
complete as it matched experience more closely.

### 1.1.4  965–1039 A.D.

The next real development in the understanding and investigation into the
nature of light came in the Middle East and Egypt. Abu Ali al-Hasan ibn al-
Haytham (also known by the Latinization Alhazen), born in Basra, Iraq, was
an active scholar in both Iraq and in Cairo, Egypt. ibn al-Haytham studied in
great detail the works by the previous ancient Greek philosophers. He used
that work as the foundation for his many research efforts with mirrors, glass
spheres, and lenses. His scholarly interests ranged from visual perception
to the nature of space itself. He asked the question "Why does the Moon
appear larger near the horizon than it does when higher up in the sky?"
His research was published in many papers as well as in his historic seven-
volume work known as the *Book of Optics*. It is in this book where he discusses

**FIGURE 1.10**
The first page of the first printed Latin translation of Alhazen's *Book of Optics*. Note how the page depicts Archimedes burning the Roman ships at the Siege of Syracuse along with other things related to light and optics.

in detail observations and experiments dealing with reflection, refraction, and lenses and mirrors. Figure 1.10 shows the first page of the first printed Latin translation of his book. It is extremely interesting to note that the page depicts Archimedes' burning of the Roman ships at the Siege of Syracuse. Apparently, al-Haytham accepted the story as fact in the 11th century.

At this point, al-Haytham was very close to developing what would eventually be known as the Snell's Law of refraction as well as the law of reflection but he didn't quite take his analyses far enough. However, in 984 A.D. the Persian scientist Ibn Sahl wrote a manuscript entitled *On Burning Mirrors and Lenses* where he developed the law and used it to describe light passing through lenses that focused light into a spot with no geometric aberrations. So, truly, from a historical perspective, perhaps Snell's Law should be known as Sahl's Law. Figure 1.11 shows the page from Sahl's manuscript where he first discovered and described this phenomenon.

### 1.1.5　1175–1294 A.D.

Moving into the 12th century, an English statesman named Robert Grosseteste (1175–1253) was beginning to develop what finally evolved into the scientific method. From about 1220 to 1235, he wrote many treatises. Some of those treatises were:

- *De sphera*. An introductory text on astronomy.
- *De luce*. On the "metaphysics of light" (which is the most original work of *cosmogony* in the Latin West).

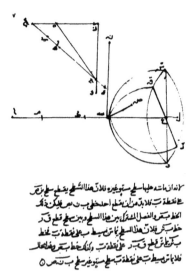

ءاندانماسده عليهاسطح مستويغيره فلان هنا الشطح يقطع سطح بزتس
عانعطة ت فلابذمن أن يطلع احدخطي ب ن بص فلكن ذلك
الخط بشر والفصل المشترك بين هذا السطح وبين سطح قطر قر
خط سر فلان هنا السطح ياترن سبط ب على نعطة ت نخط
ب كطرقطع نب ر على نعطة ت وذلك خط بشر ويفاصل
فلدايا ترن سبط ت على نعطة ت سطح مستويغيرسطح ب تنص ٥

**FIGURE 1.11**
In 984 A.D. the Persian scientist Ibn Sahl wrote a manuscript entitled *On Burning Mirrors and Lenses* where he developed the law of refraction and used it to describe light passing through lenses that focused light into a spot with no geometric aberrations. So, truly, from a historical perspective, perhaps Snell's Law should be known as Sahl's Law.

- *De accessu et recessu maris.* On tides and tidal movements (although some scholars dispute his authorship).
- *De lineis, angulis et figuris.* Mathematical reasoning in the natural sciences.
- *De iride.* On the rainbow.

In his treatise De Iride, Grosseteste wrote:

> This part of optics, when well understood, shows us how we may make things a very long distance off appear as if placed very close, and large near things appear very small, and how we may make small things placed at a distance appear any size we want, so that it may be possible for us to read the smallest letters at incredible distances, or to count sand, or seed, or any sort of minute objects.

It would seem that Grosseteste was beginning to envision a future with tools that we now know of as the telescope and microscope. The scholar, theologian, scientist, and philosopher shown in Figure 1.12 was making quite an impact on others, as can be seen of the Franciscan friar Roger Bacon.

Bacon, posthumously known by the scholastic accolade of Doctor Mirabilis, wrote a text specifically for Pope Clement IV. The text titled *Opus Majus* was printed in 1267 and included discussions of mathematics, astronomy, astrology, dynamics, optics, and alchemy. In this text, he speculates that in

**FIGURE 1.12**
Robert Grosseteste wrote many treatises on optics, light, mathematics, and astronomy and is considered to have been one of the originators of the scientific method.

the future devices could be that would improve vision. Bacon states that one day we could...

> ...shape transparent bodies, and arrange them in such a way with respect to our sight and objects of vision, that the rays will be refracted and bent in any direction we may desire, and under any angle we wish we shall see objects near or at a distance. Thus from an incredible distance we might read the smallest of letters and number grains of dust and sand.

Clearly, Bacon is envisioning optical devices like the modern-day telescope and microscope, eyeglasses, and even diffractive optics. It is also interesting to note how he describes light as "rays." In fact, he relied heavily on works of his predecessors on the topic including Ptolemy, al-Kindi, al-Haytham, and Ibn Sahl. It would seem from comparing the above quote from *Opus Majus* to the quote from *De Iride* that Robert Grosseteste was a big influence as well. The two quotes are very similar. Figure 1.13 shows a drawing from his text explaining refraction through a spherical container of water.

### 1.1.6  1300–1519 A.D.

Thomas Bradwardine was an English cleric, scholar, mathematician, physicist, and even an Archbishop of Canterbury whose life was somewhat short (1300–1349). During the so-called "Medieval Society," he was one of the bright thinkers who would help lead the European society into the Renaissance. Bradwardine was the first to suggest that light and color were

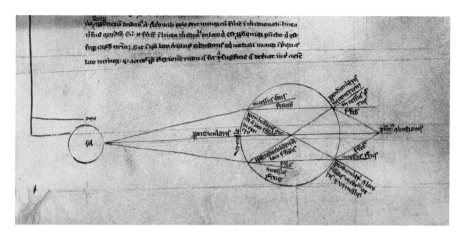

**FIGURE 1.13**
In his *Opus Majus* Roger Bacon described how "rays" of light were bent and refracted through a spherical container of water.

related. His work *Geometrica Speculativa* (see Figure 1.14) was very significant to mathematical physics and geometry, which became a much needed tool in the study of light.

Leonardo da Vinci (1452–1519), the Italian Renaissance polymath, added to the understanding of light in 1480 by making a comparison between the reflection of light and the reflection of sound waves. Though it is unclear if da Vinci was the originator of the idea or not, this is around the time that the

**FIGURE 1.14**
In his *Geometrica Speculativa* Thomas Bradwardine developed important mathematical physics and geometry needed in the further understanding of light.

**FIGURE 1.15**
Leonardo da Vinci studied light and optics in detail in order to create his masterpieces such as *The Last Supper* (1498). He was the first to compare light reflection to sound-wave reflection.

notion of light being waves or having wavelike properties appears. So at the least, he was at the cusp of new scientific ideas.

da Vinci was tedious and meticulous in his studies of light and optics in order to make the illumination within his famous artworks appear as natural and luminous as possible. He was one of the first to realize that the areas where faces and other objects were brightest must be illuminated by more light rays or were more reflective. His studies and experiments led him to the creation of such masterpieces as *The Last Supper* shown in Figure 1.15. Note that even after hundreds of years the variations of light and dark spaces are one of the qualities of this work that stands out even to the nonartist. Clearly, da Vinci's addition to humanity's understanding of light has been somewhat forgotten in history.

### 1.1.7  1550–1655 A.D.

Rene Descartes (1596–1650), shown in Figure 1.16, theorized that light was a mechanical property of a luminous body. In 1637, he published a theory describing the refraction of light where he suggested that light traveled faster in a denser medium than a less dense one. Today of course we know this to be incorrect. Perhaps he had read of da Vinci's comparison of light reflection to sound-wave reflection though the origin of his theory is unclear. However, it is known that Descartes believed that light worked like sound waves, which do indeed travel faster in the denser medium. While he was incorrect about the speed of the light in various media, he was the first to use the speed of light and media changes to describe refraction. And indeed, his notion that light "behaves" like sound waves is somewhat correct. It should

**FIGURE 1.16**
Rene Descartes first used the speed of light to describe refraction in a medium.

also be noted here that Descartes also believed that light traveled through a medium that space was filled with he called the "plenum." It should be noted here that while Descartes implied a wave nature to light he believed it to be more of a particle of this medium and is considered to be an "atomist."

Pierre Gassendi (1592–1655), shown in Figure 1.17, was a French mathematician and astronomer who was the first to measure Mercury's transit across the sun and to use a *camera obscura* to measure the diameter of the

**FIGURE 1.17**
Pierre Gassendi was a proponent of the corpuscular theory of light.

Moon. Gassendi is also, like Descartes, considered an atomist. But, unlike the ancient Greek philosophers, Descartes and Gassendi carried the idea a bit further and allowed for the tiniest particles, "atoms," to be divisible and thus became the *corpuscular theory*. The corpuscular theory suggests that all matter is composed of minute particles, which could be divided.

## 1.2 The Mathematical Era Begins

### 1.2.1 1600–1710 A.D.

Pierre de Fermat (1607–1665), a French lawyer and mathematician, shown in Figure 1.18, is important to the development of humanity's understanding of light for two main reasons. First, he made major contributions to mathematics, number theory, geometry, probability, and components of what would become calculus. His work in mathematics helped lay the groundwork for the new mathematical era that was about to begin. Second, Fermat expanded on Hero's concept that light traveled through the shortest path, but instead explained that light would travel the path that takes the least amount of time to do so. This became known as *Fermat's principle*. This was the first time it was proposed that light speed is finite and variable based on the medium it is traveling through. *Fermat's principle* paved the way to truly formulate refraction and Snell's law.

**FIGURE 1.18**
Pierre de Fermat developed the principle that light travels the path that takes the least amount of time to do so.

To truly evolve humanity's understanding of light some new math had to be created. It was about this time in the 1600s when things truly started to change with math and science. The most important tool to come along and enable humanity to truly develop theories of how the universe and light might interact with each other was the invention of calculus.

There was a controversy at the beginning of calculus as to who exactly invented it. The famous English polymath Sir Isaac Newton (see Figure 1.19) published his *Principia* in 1687 in which calculus concepts are discussed in detail. There are other smaller works and notes that suggest Newton started making his investigation of calculus, which he called the math of "fluxions and fluents" as early as 1666. It wasn't until 1704 that Newton had published his complete concept of this new math.

The German polymath Gottfried Wilhelm Leibniz (see Figure 1.20) had a breakthrough in 1675 when he first implemented the idea of using infinitesimal rectangles to determine the area under the curve of a continuous function. This was the first true application of *integral calculus*. He used the notation that we use today for integrals and differentials as shown in Equations 1.1 and 1.2.

$$\int = \text{The symbol for integration or the antiderivative.} \qquad (1.1)$$

And

$$\frac{dy}{dx} = \text{The symbol for the differentiation of } y \text{ with respect to } x. \qquad (1.2)$$

**FIGURE 1.19**
Sir Isaac Newton developed calculus and conducted many early experiments with optics.

**FIGURE 1.20**
Gottfried Wilhelm Leibniz was first to develop the antiderivative and modern notation used in calculus.

Both of the great thinkers accused the other of plagiarism but neither of their mathematical works were complete without the other and also additions that would be added by other scientists and mathematicians over the years. However, it was in this era that calculus became available to mankind and these two men are widely accepted as the co-inventors of it.

Newton moved forward with many discoveries pertaining to light during this time frame, including the invention and/or modifications to many types of telescopes, microscopes, and other optical devices. Newton was the first to conclude that color was a property of light and he demonstrated this by using white light and prisms to show the colors of the rainbow through refraction. He also demonstrated the reverse to be true as well. He first published notes called, *Of Colours*, which later with much expansion became the now famous work *Opticks* that describes light as being "corpuscular" in nature. Using the general corpuscular theory championed by Descartes and Gassendi, Newton developed the *corpuscular theory of light*.

His theory states that light is made up of tiny "corpuscles" or particles that travel in a straight line and always at a finite velocity and they possess kinetic energy. He attributed to these corpuscles the two properties that:

1. The corpuscles are perfectly weightless, rigid, and elastic.
2. All sources of light emit many of these corpuscles into some type of medium, which surrounds the source.

While Newton believed that light was made up of tiny particles he failed to be able to implement the corpuscular theory to describe and explain diffraction, interference, and polarization of light. In his *Treatise on Light*, the Dutch physicist Christiaan Huygens (1629–1695), shown in Figure 1.21, had the first true mathematical theory of light that described light as wavelike in nature. Huygens' theory, which has become known as *Huygens' principle*, stated:

> Any point on a wavefront can be considered to be a point source producing spherical secondary wavelets. The tangential surface of the secondary wavelets predict the new position of the wavefront over time.

He also pointed out that the speed of light was finite as it propagated as spherical waves. The concept of these points emitting wavelets that create a wavefront that in turn then are new points emitted along the next wavefront some small time increment later are shown in Figure 1.22.

At first, Huygens' theory wasn't accepted because it was in direct opposition of Isaac Newton who was much more renowned in the 17th-century scientific society. But simply due to the fact that Newton's theory failed to mathematically explain various aspects of the nature of light, Huygens had won out. In fact, the wave theory of light prevailed from that point up until the 20th century, but we'll get to that in a bit.

The only drawback, or at least a perceived drawback, to *Huygen's principle* was that it required the universe to be filled with a "luminiferous ether" for the wavefront to propagate through. This "ether," also spelled as "aether," was an all pervading medium much like Descartes' "plenum" that was some type of matter or fluid or other as yet undescribed material medium. Huygens described the medium required for his theory to be:

> omnipresent, perfectly elastic medium having zero density, called aether.

**FIGURE 1.21**
Christiaan Huygens developed the first truly mathematical theory of light.

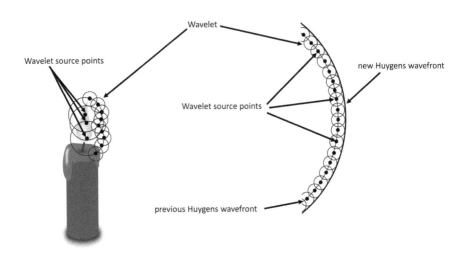

**FIGURE 1.22**
Christiaan Huygens' wavelet concept.

At the time of Huygens, the idea of the aether was perfectly acceptable. It wasn't until the early 20th century when it became a problem. Interestingly enough, in the 21st century, we now understand there is something in the nothingness of the vacuum in space between matter, but we are getting ahead of ourselves.

Another major point discussed by Huygens in his *Treatise on Light* was the observation by Ole Romer (1644–1710), a Danish astronomer shown in Figure 1.23, who through observations of Jupiter's moon Io calculated

**FIGURE 1.23**
Ole Romer was first to measure the speed of light.

the speed of light for the first time. Romer calculated light speed to be about 220,000 km/s, which is close to the now more accurate measurement of 299,792 km/s. While Romer's measurement was off by about 26% the major impact on physics of the time was proving that the speed of light was finite and not instantaneous. To this point in history, the speed of light was still conjecture and finally the conjecture was over. The speed of light was finite!

### 1.2.2  1710–1840 A.D.

But the battle of particle versus wave waged on for a century. Newton was too famous, popular, and renowned as a scholar for his particle theory to be so easily disposed of. The English physicist Thomas Young (1773–1829), shown in Figure 1.24, soon added so much experimental evidence and clarity to the wave theory of light that the community had no choice but to take it seriously. Using almost everyday commonplace items, he performed experiments that were so easy to repeat and verify that it turned the physics world upside down in regards to the theory of light.

Young's famous experiment implemented a single card a little less than a millimeter thick into a beam of sunlight shining through an open window. He then observed fringes that varied in color and had light and dark spaces in them. His most famous experiment is known as *Young's double-slit experiment*. In this experiment, he passed a beam of light through two very narrow slits that were fairly close to each other. With this experiment, he observed in great detail the diffraction of light and subsequent interference patterns created on a screen. Young surmised from his experiments that

**FIGURE 1.24**
Thomas Young showed the wave nature of light through his famous two-slit experiment.

**FIGURE 1.25**
Augustin-Jean Fresnel delivered the final blow to Newton's corpuscular theory of light.

light indeed acted as a wave, otherwise, how would such "wave-like" phenomena occur? He presented this work at the Royal Society in London in 1801 in a paper titled "On the Theory of Light and Color" and then again in 1803. More details were published in 1804 in *Philosophical Transactions* and in a subsequent paper entitled "Experiments and Calculations Relative to Physical Optics."

The double-slit experiment is elegant and became a major staple for all optical scientists to study in detail from that point to present day. In fact, aspects of light and the double-slit experiment are still debated and still throw both light and shadow on modern theory. We will circle back around to the double-slits from time to time in our quest to discern the nature of light.

In fact, there are many scientists throughout history, who used Young's double-slit to improve the details of various aspects of the theory of light. The French engineer and physicist Augustin-Jean Fresnel (1788–1827), shown in Figure 1.25, reproduced the double-slit experiment among others in order to study the wave nature of light. He expanded on *Huygen's principle*, which is now known as the *Huygens–Fresnel principle*. The work Fresnel did was the final blow to Newton's corpuscular theory and it wouldn't be until the end of the 19th century before it was reconsidered.

### 1.2.2.1 Huygens–Fresnel Principle

Up until Huygens and really until Young, most phenomena with light were considered to be due to reflection or refraction and were purely geometrical in nature. But Young's experiments showed a new thing happening and that new thing was light interacting with itself and the objects placed within its path during propagation. This is known as *diffraction*. Actually, this is truly what Huygens' wavelets and wavefront concept describes and enables us to predict. Several scientists of the time added their approaches, tricks, techniques, and insights to the development of the underlying math to the wave theory, including Green, Rayleigh, Sommerfeld, and Kirchhoff. But it was truly Fresnel's additions that enabled a new and very precise formulation to describe what happens to light as it propagates and passes by objects within its path. Let's restate the *Huygens–Fresnel principle* here as described below.

> The propagation of light is wave-like in nature and can be predicted as a wavefront, whereas each point on this wavefront acts as a point source where secondary waves are emitted. The new wavefront is the surface that envelopes all of these secondary waves.

Consider the geometric arrangement shown in Figure 1.26 where, $P_0$, is a point in the $x$, $y$ plane of interest. The $\xi, \eta$ plane is before the $x$, $y$ plane and both are situated perpendicular to the propagation axis, $z$, as shown. The point, $P_1$, is a location within the aperture opening, $\Sigma$, and is one of the Huygens–Fresnel point sources on the wavefront. The *Huygens–Fresnel principle* can be written as

$$U(P_0) = \frac{1}{j\lambda} \iint_{\Sigma} U(P_1) \frac{e^{jkr_{01}}}{r_{01}} \cos(\theta) ds. \tag{1.3}$$

In Equation 1.3, $U(P_0)$ is the light "field" which has real amplitude and complex phase components, at some distance $r_{01}$ from the Huygens–Fresnel source point, $P_1$, $U(P_1)$ is the real amplitude and complex phase function of the spherical wave there. The value $j$ is the imaginary number equal to the square root of −1. The value $k$ is known as the "wave number" and is

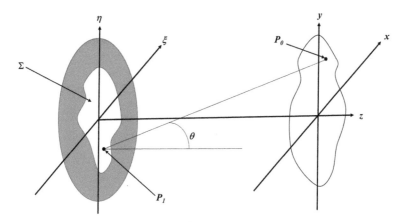

**FIGURE 1.26**
Drawing to describe Huygens–Fresnel diffraction theory.

calculated as $k = 2\pi/\lambda$ where $\lambda$ is the wavelength of the light wave. Also note that for this equation to hold true then $r_{01} \gg \lambda$.

The angle, $\theta$, is between the perpendicular normal vector to the spherical wavelet source, $\hat{n}$, and the vector, $\overrightarrow{r_{01}}$, between $P_0$ and $P_1$. Realizing that

$$\cos(\theta) = \frac{z}{r_{01}}. \tag{1.4}$$

Equation 1.3 becomes

$$U(P_0) = \frac{z}{j\lambda} \iint_\Sigma U(P_1) \frac{e^{jkr_{01}}}{r_{01}^2} ds. \tag{1.5}$$

Now inputting the appropriate plane dimensions for $P_0$ and $P_1$ and realizing that the integration is from plus to minus infinity yields the rectangular coordinate form of the *Huygens–Fresnel principle* as

$$U(x,y) = \frac{z}{j\lambda} \int_{-\infty}^{\infty}\int U(\xi,\eta) \frac{e^{jkr_{01}}}{r_{01}^2} d\xi\, d\eta. \tag{1.6}$$

Equation 1.6 is the *Huygens–Fresnel principle* in a useful form enabling the prediction of diffraction of light as it passes through an aperture and then propagates a given distance. Interestingly enough this theory has its limitations. First, what value or formulation should be used for $r_{01}$ when performing calculations? The exact value is given by

$$r_{01} = \sqrt{z^2 + (x-\xi)^2 + (y-\eta)^2}. \tag{1.7}$$

Upon inspection, it becomes clear that substituting Equation 1.7 into Equation 1.6 makes the integrals in Equation 1.6 extremely complicated and therefore approaching unusable—most certainly in the days before high-performance computers and even the calculator was invented! So, Fresnel performed an approximation on $r_{01}$. He realized that Equation 1.7 can be rewritten in the form of

$$r_{01} = \sqrt{1+b} = z\sqrt{1 + \left(\frac{x-\xi}{z}\right)^2 + \left(\frac{y-\eta}{z}\right)^2} \tag{1.8}$$

which can be rewritten through the *Taylor series* expansion

$$\sqrt{1+b} = 1 + \frac{1}{2}b - \frac{1}{8}b^2 + \cdots \tag{1.9}$$

Here is where Fresnel was clever and performed what is now known as the *Fresnel approximation*. He realized that for the $r_{01}^2$ in the denominator that he could exclude all the terms but $z$ without inducing noticeable errors in the calculation. But for the $r_{01}$ that appears in the exponent he used the second term of the binomial expansion because the exponential piece is the phase term and tiny changes of even fractions of radians in phase angles can change the calculation dramatically. Finally, the *Fresnel approximation* of the *Huygens–Fresnel principle* was found to be

$$U(x,y) = \frac{e^{jkz}}{j\lambda z} \int\!\!\!\int_{-\infty}^{\infty} U(\xi,\eta) e^{j\frac{k}{2z}\left((x-\xi)^2 + (y-\eta)^2\right)} d\xi\, d\eta. \tag{1.10}$$

Equation 1.10 is often referred to as the "Fresnel integral" or the "Fresnel diffraction integral" and it is this equation that gave humanity its first very accurate mathematical formula for describing how light behaves as it propagates. That said, it became clear fairly quickly that the *Fresnel approximation* had some problems. There is a region very close to the aperture where the equation is inaccurate and then it works. The region within which Equation 1.10 begins to become very accurate in describing light propagation is now known as the "Fresnel region" and sometimes the "Fresnel zone." Radio antenna designers typically describe the region where Fresnel diffraction is inaccurate at the "Reactive region" and the region where it is accurate as the "Radiative region." Although the Fresnel integral is accurate from some distance away from the aperture and then on for infinity it is difficult to calculate even after Fresnel's approximation approach simply because of the complexity of the calculation.

Along the same time frame the Bavarian physicist Joseph Ritter von Fraunhofer (1787–1826), shown in Figure 1.27, recognized this problem with the Fresnel approximation and created an approximation of his own that made the equation much easier to calculate. Fraunhofer rewrote the equation as

$$U(x,y) = \frac{e^{jkz}}{j\lambda z} \int\!\!\!\int_{-\infty}^{\infty} U(\xi,\eta) e^{j\frac{k}{2z}\left(x^2 + y^2\right)} e^{-j\frac{k}{z}(x\xi + y\eta)} e^{j\frac{k}{2z}\left(\xi^2 + \eta^2\right)} d\xi\, d\eta. \tag{1.11}$$

He then realized that for

$$z \gg \frac{k}{2}\left(\xi^2 + \eta^2\right) \tag{1.12}$$

then this particular quadratic phase factor was approximately one over the entire aperture. Finally, the *Fraunhofer approximation* became

$$U(x,y) = \frac{e^{jkz} e^{j\frac{k}{2z}\left(x^2 + y^2\right)}}{j\lambda z} \int\!\!\!\int_{-\infty}^{\infty} U(\xi,\eta) e^{-j\frac{k}{z}(x\xi + y\eta)} d\xi\, d\eta. \tag{1.13}$$

**FIGURE 1.27**
Joseph Ritter von Fraunhofer recognized the problem with the Fresnel's approximation and created an approximation of his own that is still used today to model diffraction of optical wavefronts.

Equation 1.13 is much simpler to calculate than Equation 1.10; however, it is also limited even more stringently. The *Fraunhofer approximation* doesn't become accurate until the region known as the "far field." A widely published "rule of thumb" known as the "antenna designer's formula" suggests that the *Fraunhofer approximation* is only accurate at distances of

$$z > \frac{2D^2}{\lambda} \tag{1.14}$$

where $D$ is the diameter of the aperture. Figure 1.28 shows the regions where the Fresnel and Fraunhofer approaches should be used. Note that in the region very close to the aperture a much more complex approach must be used to calculate the light field propagation and is often referred to as the "Rayleigh region." The location at which to transition from the Rayleigh region to the Fresnel region is a bit ambiguous and treated differently based on the reference one checks. The general rule where the Fresnel integral is useful is for the propagation distance to be much greater than the wavelength of the light. The problem with this definition is that the start of the Fresnel region is a bit arbitrary. Just what does the wavelength being much smaller than the propagation mean? How "much smaller" is small enough? Another general "antenna designer's rule of thumb" is that the Fresnel region can be considered to be between

$$0.62\sqrt{\frac{D^3}{\lambda}} < z > \frac{2D^2}{\lambda}. \tag{1.15}$$

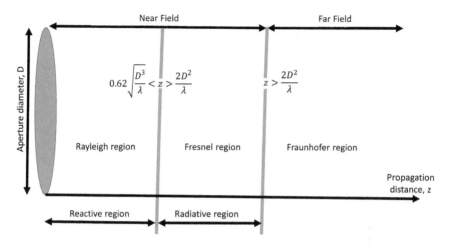

**FIGURE 1.28**
Fresnel and Fraunhofer developed methods for calculating light propagation in the near-field and far-field regions.

At this point, we must point out that the integral in Equation 1.13 describes the spatial distribution of the so-called "light field." There are no instruments to implement to allow us to understand truly, what that might mean. Modern-day detection methods actually measure the irradiance, $I(x, y)$, of the field, which is all real, measured in Watts per meter squared, and is found by

$$I(x,y) = U^*(x,y)U(x,y) = |U(x,y)|^2. \qquad (1.16)$$

The superscript * denotes the complex conjugate. Now is as good a time as any to point out something about the quantity, $I$. As stated above the units of measurement for this quantity are watts per square meter and it is called the irradiance, which is defined as the radiant flux or power received by or incident upon a surface per unit area—hence, the power per area-type units. Some radiometry textbooks use the letter $E$ rather than $I$ for irradiance, but then that becomes confusing with the symbol for the electric field, which is $E$. So, in those books they use a different font to distinguish between the two.

Many textbooks use other names for this quantity and therefore become confusing because of the different nomenclature. Most optics books will refer to this quantity as "intensity" or more accurately with a descriptor as "optical intensity." That is quite confusing because the intensity is also defined as a measure of the power emitted, reflected, transmitted, or received per unit solid angle rather than area. In astronomy, the quantity is often referred to as "apparent brightness" and is represented by, $b$, and sometimes, though rarely, even, *F*. There are other nomenclatures that can add further to this confusion and we will see in the next section the Poynting vector, **S**, which is the power per unit area of the electromagnetic field.

The point of this obfuscation is to realize that we need to set a standard for this text and keep it that way throughout. It should also stand out to the reader that the key thing to determine with any of these nomenclatures is the units of measurement. Watts per square meter is clearly a measure of power per unit area no matter what it is called or what symbol is used to represent it! But herein lies the conundrum. Properly using irradiance would require the use of *E*, which we will also want to use as the electric field later and maybe even for energy giving us the need for three different fonts for the same letter representing three different quantities. That would certainly be confusing. Or, we could say, "optical intensity" every time and use *I*, and therefore risk being accused of being sloppy with our nomenclature for as we discussed above "intensity" means something else. Heretofore, in order to avoid further obfuscation, this text will use the letter *I* in Cambria Math font to represent the power per unit area measurement of the light field as defined in Equation 1.16 and we will call it "irradiance."

### 1.2.2.2 The Fourier Transform

Before we can move on with our story of the history of our understanding of light, we need to realize something quite fantastic about the Fresnel and Fraunhofer integrals shown in Equations 1.10 and 1.13. They can be rewritten into a much more useful form that allows for solving many complex calculations of light fields passing by various geometrically shaped apertures almost from inspection (with some practice, lots of practice). The two equations can be rewritten as some multiplicative factors and phase terms multiplied by the *Fourier Transform* of the aperture geometry or the field's spatial geometry in the aperture plane.

About the same time frame as Fresnel and Fraunhofer, Joseph Fourier (1768–1830), shown in Figure 1.29, was a French mathematician who developed math that would eventually be called the *Fourier Transform* in his honor. It will be assumed that the reader has been exposed to Fourier's math at this point so a development of it will not be given here. However, application of it is very useful when it comes to discussing light and laser beams. So, we need to at least become acquainted with some of its applications.

For our purposes, we will limit ourselves to the "far field" and the Fraunhofer integral. Equation 1.13 can be rewritten as

$$U(x,y) = \frac{e^{jkz}e^{j\frac{k}{2z}(x^2+y^2)}}{j\lambda z} \mathcal{F}\{U(\xi,\eta)\} \tag{1.17}$$

where $\mathcal{F}$ represents the Fourier Transform.

**FIGURE 1.29**
Joseph Fourier was a French mathematician who developed math that would eventually be called the *Fourier Transform*.

### Example 1.1: The Propagation of Light through a Rectangular Aperture

Consider the rectangular aperture as shown in Figure 1.30. This "rectangular aperture" is physically nothing more than a rectangular-shaped hole in an opaque screen. The rectangular aperture has dimensions of being $a$ wide in the $x$ dimension and $b$ tall in the $y$ dimension. We describe this aperture as the two-dimensional "rectangle function" or the "rect function" and it is written as the multiplication of two one-dimensional functions. The rect function is

$$\text{rect}\left(\frac{x}{a}\right) = \begin{cases} 1 & x < \dfrac{a}{2} \\ \dfrac{1}{2} & x = \dfrac{a}{2} \\ 0 & \text{otherwise} \end{cases} \tag{1.18}$$

The two-dimensional function, $U(\xi, \eta)$, describing the light field immediately following the aperture is

$$U(\xi, \eta) = \text{rect}\left(\frac{\xi}{a}\right)\text{rect}\left(\frac{\eta}{b}\right). \tag{1.19}$$

Note that this can also be written as

$$U(\xi, \eta) = \text{rect}\left(\frac{\xi}{a}\right)\text{rect}\left(\frac{\eta}{b}\right) = \text{rect}\left(\frac{\xi}{a}, \frac{\eta}{b}\right). \tag{1.20}$$

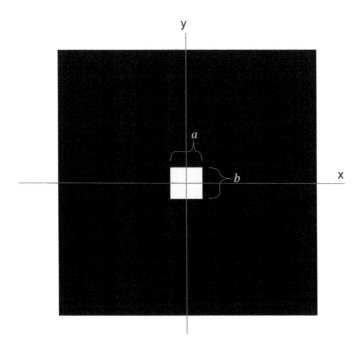

**FIGURE 1.30**
The rectangle function describes a rectangular opening in an opaque screen.

Substituting Equation 1.20 into Equation 1.17 allows us to calculate the light field spatial distribution in the far field as

$$U(x,y) = \frac{e^{jkz}e^{j\frac{k}{2z}(x^2+y^2)}}{j\lambda z} \mathcal{F}\left\{\text{rect}\left(\frac{\xi}{a},\frac{\eta}{b}\right)\right\}. \qquad (1.21)$$

Table 1.1 gives some functions and their Fourier Transform pairs. From Table 1.1 we see that the Fourier Transform of the rect function is the so-called "sinc" function (or $\sin(x)/x$ which is pronounced like the kitchen "sink"). Performing the Fourier Transform on Equation 1.21 we will go from spatial dimensions, $(\xi,\eta)$, near the aperture plane all the way out to the spatial frequency dimensions, $(f_x, f_y)$ in the far field. The light in the far field is then

$$U(x,y) = \frac{e^{jkz}e^{j\frac{k}{2z}(x^2+y^2)}}{j\lambda z} ab \, \text{sinc}\left(af_x, bf_y\right). \qquad (1.22)$$

In order to express the light field in spatial dimensions rather than spatial frequency dimensions we must realize that $f_x = x/\lambda z$ and $f_y = y/\lambda z$. Therefore,

**TABLE 1.1**

Some Useful Fourier Transforms

$$\mathcal{F}\{1\} \rightarrow \delta(f_x, f_y)$$

$$\mathcal{F}\{\delta(ax, by)\} \rightarrow \frac{1}{|ab|}$$

$$\mathcal{F}\{e^{j\pi(ax+by)}\} \rightarrow \delta\left(f_x - \frac{a}{2}, f_y - \frac{b}{2}\right)$$

$$\mathcal{F}\{\cos(ax)\} \rightarrow \pi\left[\delta(f_x - a) + \delta(f_x + a)\right]$$

$$\mathcal{F}\{\sin(ax)\} \rightarrow j\pi\left[\delta(f_x + a) - \delta(f_x - a)\right]$$

$$\mathcal{F}\left\{\text{rect}\left(\frac{\xi}{a}, \frac{\eta}{b}\right)\right\} \rightarrow ab\,\text{sinc}(af_x, bf_y)$$

Equation 1.22 is the light field in the far field according to the Fraunhofer diffraction integral. Note that the field is complex and is not something that is particularly measurable. So, in order to calculate what a detector such as a photograph or a digital camera might detect, we must now calculate the irradiance pattern in the far field using Equation 1.16. So,

$$I(x,y) = U^*(x,y)U(x,y) = |U(x,y)|^2 \tag{1.23}$$

$$I(x,y) = \frac{e^{-jkz}e^{-j\frac{k}{2z}(x^2+y^2)}}{-j\lambda z}ab\,\text{sinc}\left(a\frac{x}{\lambda z}, b\frac{y}{\lambda z}\right)$$

$$\frac{e^{jkz}e^{j\frac{k}{2z}(x^2+y^2)}}{j\lambda z}ab\,\text{sinc}\left(a\frac{x}{\lambda z}, b\frac{y}{\lambda z}\right) \tag{1.24}$$

$$I(x,y) = \frac{a^2 b^2}{\lambda^2 z^2}\text{sinc}^2\left(a\frac{x}{\lambda z}, b\frac{y}{\lambda z}\right). \tag{1.25}$$

Equation 1.25 is the irradiance pattern that would be detected by a camera in the far field and is shown in Figure 1.31. It should be noted here that the far field might be at some large propagation distance as described in Equation 1.14; however, it can also be reached by passing the light field through a convex lens or bouncing the field of a concave mirror and forcing the light field to focus into a "focal spot." This focal spot is also sometimes referred to as the "Fourier plane" or "focal plane" and will represent the far-field spatial distribution. As shown in Figure 1.32, propagation of the light field to the far field can also be by focusing the light field with a lens or mirror. This characteristic of light,

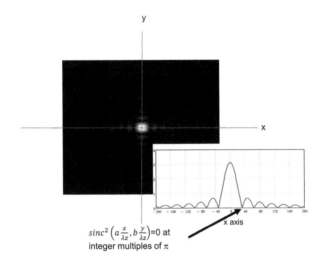

$sinc^2\left(a\frac{x}{\lambda z}, b\frac{y}{\lambda z}\right)=0$ at
integer multiples of $\pi$

**FIGURE 1.31**
The Fraunhofer diffraction pattern of a rectangular aperture is a sinc² function.

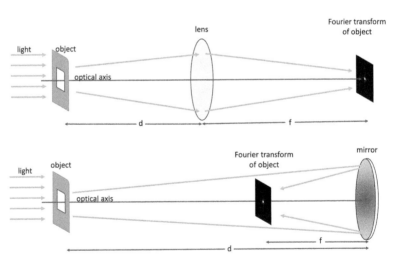

**FIGURE 1.32**
Lenses and mirrors can be used to propagate a beam to the Fourier transform plane.

lenses, and mirrors makes laboratory experiments much simpler than always requiring propagation to great distances to reach the far field and therefore make measurements of them. Also, note that there are many modern-day laser designs that for mechanical construction purposes might require having a rectangular aperture or having rectangular optics and we shall note here that this calculation in Example 1.1 will be quite useful throughout our discussion on lasers.

### 1.2.2.3 Parseval's Theorem

Marc-Antoine Parseval (1755–1836) was a French mathematician of the same era as Fresnel, Fraunhofer, and Fourier. Parseval made a very important discovery that was in essence a restatement of the law of conservation of energy. The law of conservation of energy states that energy can neither be created nor destroyed, must be conserved over time, and therefore is merely transformed from one form to another. Using the Fraunhofer diffraction as shown in Example 1.1, we see that a light field incident on a rectangular aperture goes from taking the spatial shape of a rectangle in $x$ and $y$ in the aperture plane to that of the sinc$^2$ function spread out across an arbitrary plane in the far field. Parseval realized that if somehow all of the energy in the rectangle weas measured and compared to a measurement of all the energy in the far field plane, the two would be the same. But he didn't do this through experiment. He determined this simply due to the consequences of the mathematics of performing a transform on a function. Parseval's theorem states if

$$\mathcal{F}\{g(x,y)\} = G(f_x, f_y) \tag{1.26}$$

Then

$$\int\int_{-\infty}^{\infty}|g(x,y)|^2 \, dx \, dy = \int\int_{-\infty}^{\infty}|G(f_x,f_y)|^2 \, df_x \, df_y. \tag{1.27}$$

Parseval's theorem (sometimes also known as Rayleigh's theorem) is a very elegant mathematical formulation that tells us something very important about light. It tells us that light follows the rules of energy which, in turn, allows us to make an inference that light itself must be some form of energy. Through diffraction, the energy making up light is neither created nor destroyed. Instead, it is merely transformed from one spatial pattern to another.

### 1.2.3 1840–1899 A.D.

### 1.2.3.1 The Hankel Transform

In Example 1.1, we saw how we could use a Fourier transform to calculate diffraction effects due to a two-dimensional aperture. Fourier transform pairs will work well for such calculations as long as the functions are separable in $x$ and $y$ as is the rectangular function. In other words,

$$f(x,y) = g(x)h(y). \tag{1.28}$$

But what happens in the case where the function is not separable into dimension-specific equations? The answer is that we must use a coordinate transformation to simplify the situation.

The German mathematician, Hermann Hankel (1839–1873), shown in Figure 1.33, discovered a way to express any arbitrary function as the weighted sum of an infinite number of *Bessel functions of the first kind*, $J_v(kr)$, where the subscript $v$ is the order or type of Bessel function and in the series solution must be of the same order, $k$ is a scaling factor, and $r$ is the $r$ axis variable in circular coordinates. The Hankel transform is given as

$$F_v(r) = \mathcal{H}\{f(r)\} = \int_0^\infty f(r)J_v(kr)r\,dr. \tag{1.29}$$

It is beyond the scope of this text to go into the details of Bessel functions and how to calculate them. As with the Fourier transform, we will simply use lookup tables as the method for calculation. It would be beneficial for the reader to acquire knowledge of this level of mathematics in the future as the use of Fourier transforms and Bessel functions in science and engineering runs amok!

Using Hankel's approach, which is also sometimes known as the *Fourier–Bessel transform*, enables the calculation of Fraunhofer diffraction due to complex apertures that can be expressed in circular coordinates. The importance of this mathematical tool will become apparent the deeper we look into lasers, laser construction, and laser beam propagation because the majority of laser beams are transmitted through circular apertures. So, let's practice using the *Hankel transform*.

**FIGURE 1.33**
Hermann Hankel discovered a way to express any arbitrary function as the weighted sum of an infinite number of *Bessel functions of the first kind*.

**Example 1.2: The Propagation of Light through a Circular Aperture**

Consider a circular aperture as shown in Figure 1.34. In essence, the "circular aperture" is merely a circular hole of some radius, $R$, in an opaque screen. We will write an equation to describe this aperture plane as the "circle function" given by

$$\text{circ}\left(\frac{\sqrt{x^2+y^2}}{R}\right) = \begin{cases} 1 & \sqrt{x^2+y^2} < R \\ \frac{1}{2} & \sqrt{x^2+y^2} = R \\ 0 & \text{otherwise} \end{cases} \tag{1.30}$$

As we can see from Equation 1.30 the circular aperture function is not separable in $x$ and $y$ and therefore cannot be used in a standard *Fourier transform*. First, we must rewrite it in circular coordinates as

$$\text{circ}\left(\frac{r}{R}\right) = \begin{cases} 1 & r < R \\ \frac{1}{2} & r = R \\ 0 & \text{otherwise} \end{cases} \tag{1.31}$$

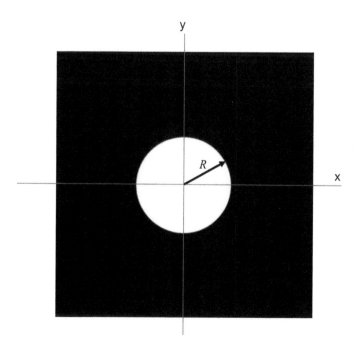

**FIGURE 1.34**
Circular aperture for Example 1.2.

where $r = \sqrt{x^2 + y^2}$ and is the radial dimension in circular coordinates. As in Example 1.1, the light field immediately following the aperture is written as

$$U(q) = \mathrm{circ}\left(\frac{q}{R}\right). \tag{1.32}$$

In Equation 1.32, we use the coordinate, $q$, to represent the radial spatial dimension in the aperture plane in the same way we used $(\xi, \eta)$ for $(x, y)$ coordinates with the rectangular aperture and is

$$q = \sqrt{\xi^2 + \eta^2}. \tag{1.33}$$

At this point, in order to calculate the Fraunhofer integral we must use the *Hankel transform*

$$U(r) = \frac{e^{jkz}}{j\lambda z} e^{j\frac{kr^2}{2z}} \mathcal{H}\left\{\mathrm{circ}\left(\frac{q}{R}\right)\right\}. \tag{1.34}$$

The *Hankel transform* of the circle function is

$$\mathcal{H}\left\{\mathrm{circ}\left(\frac{q}{R}\right)\right\} = \pi R^2 \frac{J_1(2\pi R\rho)}{\pi R\rho}. \tag{1.35}$$

Again, like in the rectangular case in Example 1.1 we are transforming from spatial dimension in the aperture plan to spatial frequency in the far-field plane. In this case, we are transforming from radial spatial dimension, $q$, to radial spatial frequency, $\rho$, where $\rho = \sqrt{f_x^2 + f_y^2} = r/\lambda z$. Now we can rewrite Equation 1.34 to determine the light field in the far field as

$$U(r) = \frac{e^{jkz}}{j\lambda z} e^{j\frac{kr^2}{2z}} \pi R^2 \frac{J_1\left(2\pi R \frac{r}{\lambda z}\right)}{\pi R \frac{r}{\lambda z}}. \tag{1.36}$$

Realizing that $k = 2\pi/\lambda$ yields

$$U(r) = \frac{e^{jkz}}{j\lambda z} e^{j\frac{kr^2}{2z}} \pi R^2 2 \frac{J_1\left(kR\frac{r}{z}\right)}{kR\frac{r}{z}}. \tag{1.37}$$

The irradiance is therefore

$$I(r) = \left(\frac{\pi R^2}{\lambda z}\right)^2 \left[2\frac{J_1\left(kR\frac{r}{z}\right)}{kR\frac{r}{z}}\right]^2.$$  (1.38)

Equation 1.38 is a very important equation when it comes to laser science and engineering, as we will see as we become more educated on the subject. The equation is known as the "Bessel-sinc squared function" sometimes the "Besinc squared" and it is also often called the *Airy pattern* after G.B. Airy the person who first derived it. Figure 1.35 shows this Besinc-squared function as well as a cross sectional plot along the radius dimension. Notice how the pattern is similar to the sinc-squared function shown in Example 1.1 on this function is radially symmetric. A very important aspect of the Besinc-squared function with regard to lasers is where the first zero occurs at

$$r = 1.22\frac{\lambda z}{R}.$$  (1.39)

Equation 1.39 tells us something very important about a light field passing through a circular aperture. What it tells us is that the radius of the beam in the far field is directly proportional to the distance it travels and inversely proportional to the radius of the aperture itself. Physically what this means is that the bigger the aperture you have, the smaller the beam you will have in the far field. Equation 1.39 shows us that a laser

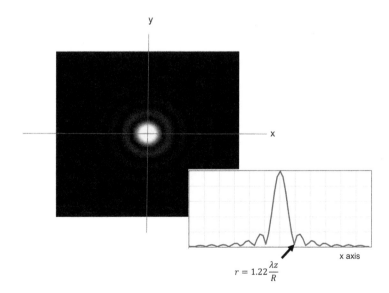

**FIGURE 1.35**
Fraunhofer diffraction of the circular aperture in the far field.

beam doesn't stay in a perfect pencil beam out to infinity as one may think. In fact, the beam spreads based on the wavelength of the light (which we will discuss soon), the distance traveled, and the size of the aperture of the laser device.

### 1.2.3.2 Maxwell's Equations

While the *Huygens–Fresnel principle*, and therefore the wave theory, for light propagation had been worked out in fairly good detail and the works of Fresnel and Fraunhofer, among others, had enabled the detailed calculations and predictions of light propagations through and past apertures and objects, it was still a bit unclear exactly what light is. While the scientific community of the early to mid-1800s had now accepted that light was a wave and they knew that when focused with lenses light would burn, so therefore, light had kinetic energy somehow. Parseval's theorem also showed us that light obeyed the law of conservation of energy, but the scientists of the era still had no true idea of what light might actually be (not that we truly do today either but we do know more now than they did back then). All that was about to change dramatically!

In 1865, the Scottish physicist James Clerk Maxwell (1831–1879), as shown in Figure 1.36, published "A Dynamical Theory of the Electromagnetic Field" in which for the first time ever he theorized that electricity, magnetism, and light while all seemingly different physical phenomena were, in fact, manifestations of the same thing. For the first time, there was a theory that

**FIGURE 1.36**
James Clerk Maxwell published "A Dynamical Theory of the Electromagnetic Field" in which for the first time ever he theorized that electricity, magnetism, and light while all seemingly different physical phenomena were, in fact, manifestations of the same thing.

explained light through other physically observed phenomena that could actually be tested through experimentation. Maxwell's seminal publication is considered to be one of the most important collections and integration of experimental observations, mathematical calculations and formulations, and unifying works in the history of physics tying electricity, magnetism, and light all into one elegant, though complex, mathematical theory.

Maxwell's work was so encompassing that it showed the connectivity between the works of Coulomb, Orsted, Gauss, Biot, Savart, Ampere, and Faraday all in one paper. In their original forms, there were some 20 different equations. At the time, the formalism of vector calculus had yet to be completed and the available mathematics needed to actually catch up with Maxwell's brilliance. Using various vector math techniques British physicist Oliver Heaviside (Figure 1.37) and American physicist Josiah Willard Gibbs (Figure 1.38) concurrently developed vector calculus and reformulated the work completed in Maxwell's research from the 20 equations down to the 4 that are more widely recognized today as "Maxwell's equations." The German physicist Heinrich Rudolf Hertz (Figure 1.39) devised experiments that verified that electromagnetic waves as predicted by Maxwell propagated at the speed of light. Hertz's work would bring forth the era of the radio in the next few decades to come.

Equations 1.40–1.43 are the modern-day accepted and so-called Maxwell's equations. Each of the equations was first discovered as individual electrical or magnetic phenomena by the scientist listed beside them and was mostly unrelated until Maxwell showed the connection between them.

**FIGURE 1.37**
British physicist Oliver Heaviside helped develop vector calculus and to reduce Maxwell's equations from many to four.

**FIGURE 1.38**
American physicist Josiah Willard Gibbs helped develop vector calculus and to reduce
Maxwell's equations from many to four.

**FIGURE 1.39**
The German physicist Heinrich Rudolf Hertz devised experiments that verified that electromagnetic waves as predicted by Maxwell propagated at the speed of light.

$$\nabla \times \bar{E} = -\frac{\partial \bar{B}}{\partial t} \qquad \left(\text{Faraday's Law}\right) \qquad (1.40)$$

$$\nabla \times \bar{B} = \mu_0 \bar{J} + \mu_0 \varepsilon_0 \frac{\partial \bar{E}}{\partial t} \qquad \left(\text{Ampere's Law}\right) \qquad (1.41)$$

$$\nabla \cdot \bar{E} = \frac{\rho}{\varepsilon_0} \qquad \left(\text{Coulomb's Law of electrostatics}\right) \qquad (1.42)$$

$$\nabla \cdot \bar{B} = 0 \qquad \left(\text{Coulomb's Law of magnetostatics}\right) \qquad (1.43)$$

In Equations 1.40–1.43, $\bar{E}$ is the electric field vector measured in volts per meter, $\bar{B}$ is the magnetic field vector measured in tesla (or newtons per meter per ampere), $\bar{J}$ is the current density vector measured in amperes per square meter (and can be found by dividing the current, $I$, by the cross-sectional area of the material, $A$), $\mu_0$ is the magnetic permeability of space measured in henries per meter (or newtons per square ampere), and $\varepsilon_0$ is the electric permittivity of space measured in farads per meter (or seconds per ohm, seconds-squared per henrie, or ampere second per volt).

But what was it about these equations that led Maxwell to the conclusion that they actually had anything to do with the phenomenon of light? In order to make that connection there are a couple of things we have to learn first. So now we must digress into, believe it or not, electrical circuit theory.

Equation 1.41 is what we know of today as Ampere's Law; however, originally it was written as

$$\nabla \times \bar{B} = \mu_0 \bar{J}. \tag{1.44}$$

Note the difference between Equations 1.41 and 1.44. Equation 1.44 says something fairly simple about current flowing through a medium. In essence, the original equation stated that (on the right-hand side of Equation 1.44) current flowing through some medium, such as a wire, would in return create a magnetic field about it (the left-hand side of Equation 1.44). And of course the reciprocal statement is true as well in that a varying magnetic field about a wire or medium will have a corresponding current flow through said medium associated with it. But there is a problem with Equation 1.44 based on experimental evidence.

Consider the circuit shown in Figure 1.40 of a capacitor connected to a power source. Per Equation 1.44, the current flowing through the wires has an associated magnetic field with it. Now examine the capacitor more closely. Assuming there is nothing but vacuum or empty space between the plates of the capacitor then no current can flow from one plate to the other; however, current flows into one plate and then out of the other plate. How can that be? Also note that there is still a magnetic field measured, as if there is a continuously flowing current there while there is none!

Maxwell realized that there was a flaw in Ampere's original circuital law and that it needed a correction. He surmised that even though there was no measureable "real" current flowing between these capacitor plates that there must be some phenomenon like a current flow that was taking place in the empty space between them. In his 1861 paper "On Physical Lines of Force" he said that this "…displacement…is commencement of a current" and from that point on has been known as the *displacement current, $\bar{J}_D$*, and is written as

$$\bar{J}_D = \varepsilon_0 \frac{\partial \bar{E}}{\partial t}. \tag{1.45}$$

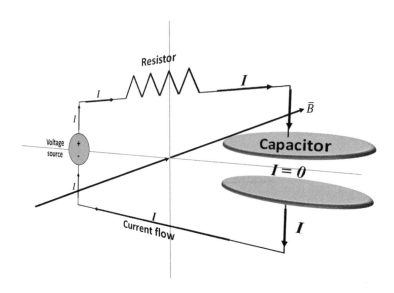

**FIGURE 1.40**
A typical electrical circuit with current flowing through it causes a magnetic field even with the open-circuit element of the capacitor.

By adding this component to Equation 1.44, Maxwell corrected Ampere's Law to match more precisely the experimental measurements, but it also enabled another major accomplishment. Using Equations 1.40–1.43 Maxwell was able to now derive from them a wave equation to describe the electric and magnetic phenomena (we now use the unified nomenclature of "electromagnetism").

### 1.2.3.3 The Wave Equation

When it comes down to electromagnetism a scientist might ask which is the more important field, quantity of electricity or magnetism? Both of the quantities are now considered as fundamental and one is not more or less important than the other. They are also each interrelated with the other in terms of how they interact with matter and can be described further through the two "constitutive equations."

$$\bar{D} = \varepsilon_r \varepsilon_0 \bar{E} + \bar{P} = \epsilon \bar{E} + \bar{P} \tag{1.46}$$

and

$$\bar{B} = \mu_r \mu_0 \bar{H} + \bar{M} = \mu \bar{H} + \bar{M} \tag{1.47}$$

where $\varepsilon_r$ is a relative multiplier for the permittivity and $\mu_r$ is a relative multiplier for the permeability of a given material medium. For free space

$\varepsilon_r = \mu_r = 1$. Also note that $\epsilon$ and $\mu$ are the overall measureable permittivity and permeability, respectively, of a given medium. $\bar{D}$ is the electric displacement field measured in coulombs per square meter. $\bar{H}$ is the "magnetic H-field" and is measured in units of amperes per meter. $\bar{P}$ is the electric field polarization vector and $\bar{M}$ is the magnetization vector. In free space Equations 1.46 and 1.47 become

$$\bar{D} = \varepsilon_0 \bar{E} \tag{1.48}$$

And

$$\bar{B} = \mu_0 \bar{H}. \tag{1.49}$$

Since the current density in free space is zero, Maxwell's equations become

$$\nabla \times \bar{E} = -\frac{\partial \bar{B}}{\partial t} \tag{1.50}$$

$$\nabla \times \bar{B} = \mu_0 \varepsilon_0 \frac{\partial \bar{E}}{\partial t} \tag{1.51}$$

$$\nabla \cdot \bar{E} = \frac{\rho}{\varepsilon_0} \tag{1.52}$$

$$\nabla \cdot \bar{B} = 0. \tag{1.53}$$

Taking a closer look at Equations 1.50–1.53 we realize that we have a set of coupled, first-order, partial differential equations of the electric and magnetic fields. In order to decouple the equations we need to make use of the vector calculus identity

$$\nabla \times \nabla \times \bar{F} = \nabla(\nabla \cdot \bar{F}) - \nabla^2 \bar{F}. \tag{1.54}$$

Apply the identity in Equation 1.54 into 1.50

$$\nabla \times \nabla \times \bar{E} = \nabla(\nabla \cdot \bar{E}) - \nabla^2 \bar{E} = \nabla \times \left( -\frac{\partial \bar{B}}{\partial t} \right). \tag{1.55}$$

Substitute the right-hand side of Equation 1.52 into the left-hand side and the right-hand side of Equation 1.51 into the right-hand side to get

$$\nabla \left( \frac{\rho}{\varepsilon_0} \right) - \nabla^2 \bar{E} = -\frac{\partial}{\partial t} \left( \mu_0 \varepsilon_0 \frac{\partial \bar{E}}{\partial t} \right). \tag{1.56}$$

Since $\rho/\varepsilon_0$ is a constant, the derivative of it is zero so that portion goes away. Simplifying and cancelling the minus signs gives the resulting equation

$$\nabla^2 \overline{E} = \mu_0 \varepsilon_0 \frac{\partial^2 \overline{E}}{\partial t^2}. \tag{1.57}$$

The same process can be used to achieve the similar decoupled magnetic field equation

$$\nabla^2 \overline{B} = \mu_0 \varepsilon_0 \frac{\partial^2 \overline{B}}{\partial t^2}. \tag{1.58}$$

At this point, we must realize that Equations 1.57 and 1.58 look very familiar and look a lot like

$$\nabla^2 \overline{f} = \frac{1}{v^2} \frac{\partial^2 \overline{f}}{\partial t^2}. \tag{1.59}$$

Equation 1.59 is the equation for a transverse wave propagating on a vibrating string with velocity, $v$. Comparing Equations 1.57 and 1.58 to Equation 1.59 we see that they can be rewritten as

$$\nabla^2 \overline{E} = \frac{1}{c^2} \frac{\partial^2 \overline{E}}{\partial t^2}. \tag{1.60}$$

and

$$\nabla^2 \overline{B} = \frac{1}{c^2} \frac{\partial^2 \overline{B}}{\partial t^2}. \tag{1.61}$$

We substituted $c$ for $v$ as our wave velocity and therefore

$$c = \frac{1}{\sqrt{\mu_0 \varepsilon_0}}. \tag{1.62}$$

Using the measured values of $\mu_0 = 4\pi \times 10^{-7}\,\text{N/A}^2$ and $\varepsilon_0 = 8.85 \times 10^{-12}\,\text{C}^2/\text{Nm}^2$ Maxwell calculated the speed of the coupled electric and magnetic waves to be

$$c = \frac{1}{\sqrt{\mu_0 \varepsilon_0}} = 2.997 \times 10^8\,\text{m/s}^2. \tag{1.63}$$

The velocity of this wave turned out to be extremely close to the measured speed of light and therefore Maxwell concluded that light must be just another form of this electric and magnetic coupled field wave! For the first time, there was actually some physics-based mathematics that told humanity what light might actually be, or at least a means of describing it in more detail. It is very interesting to note here that without Maxwell's correction to Ampere's Law the above process would not be possible. Although the nature of the displacement current seems unclear and perhaps strange, the mathematical adjustment to Ampere's Law is necessary to explain observed phenomena and in turn enable a wave description for light.

Figure 1.41 shows the general idea behind Maxwell's coupled electric and magnetic waves. We will call them *electromagnetic waves* from here on. The premise is that the fields are each transverse waves orthogonal to each other and in turn each inducing the other in order to propagate. Each of the fields can be written as *plane waves* propagating in the z direction as follows

$$\bar{E}(z,t) = E_0 \hat{p}_E e^{j(kz-\omega t)}. \tag{1.64}$$

And

$$\bar{B}(z,t) = B_0 \hat{p}_B e^{j(kz-\omega t)}. \tag{1.65}$$

In Equation 1.64, the $E_0$ is the real magnitude of the electric field, $\hat{p}_E$ is the polarization vector for the field, $k = 2\pi/\lambda$ is the *wave number* which is a function of the wavelength, $\lambda$, and $\omega$ is the angular frequency of the wave and is found by

$$\omega = ck = c2\pi/\lambda = 2\pi\nu. \tag{1.66}$$

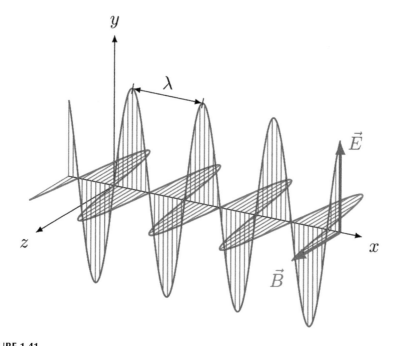

**FIGURE 1.41**
Maxwell's coupled electric and magnetic waves. Image is public domain from Wikimedia Commons.

$\nu$ is the frequency or number of times the wave peaks per second measured in hertz (note that often times frequency is also represented by $f$). It should be pointed out here that these plane waves exist at a single wavelength and frequency and therefore are called *monochromatic plane waves* (monochromatic suggests just one color and the wavelength of light is associated with its color, i.e., red light of a laser pointer has a wavelength of about 633 nm).

The only difference in Equation 1.65 is that it represents the magnetic field wave and has a magnitude, $B_0$, and a polarization vector, $\hat{p}_B$. Equations 1.64 and 1.65 can be rewritten with one polarization vector but the wave number must then become a propagation or wave vector rather than the simple scalar as shown above. So, the more general vector form of the two equations in rectangular coordinates is

$$\overline{E}(\overline{r},t) = E_0 e^{j(\overline{k}\cdot\overline{r}-\omega t)}\hat{p} \tag{1.67}$$

and

$$\overline{B}(\overline{r},t) = B_0 e^{j(\overline{k}\cdot\overline{r}-\omega t)}(\overline{k}\times\hat{p}). \tag{1.68}$$

The real amplitudes of the electric and magnetic field waves are

$$B_0 = \frac{k}{\omega}E_0 = \frac{1}{c}E_0. \tag{1.69}$$

So, Equation 1.68 can be rewritten completely in terms of the electric field wave

$$\overline{B}(\overline{r},t) = \frac{1}{c}E_0 e^{j(\overline{k}\cdot\overline{r}-\omega t)}(\overline{k}\times\hat{p}) = \frac{1}{c}\overline{k}\times\overline{E}. \tag{1.70}$$

### 1.2.3.4 The Poynting Vector

We are almost complete with the development of the classical mathematical wave theory of light. We have seen how Huygens and Fresnel, and later Fraunhofer, Fourier, and others, developed mathematical formulations of the "light field" to describe propagation and diffraction effects. We have seen how Maxwell unified electricity, magnetism, and light into four brilliant and elegant vector equations. But we need to do one final thing and that is to tie the measurable quantity of the Huygens–Fresnel wave theory, irradiance as shown in Equation 1.16, to Maxwell's electromagnetic propagating plane waves in Equations 1.67 and 1.70.

The connection between these two theories was developed by English physicist John Henry Poynting (1852–1914), shown in Figure 1.42, in a paper he authored in 1884 titled "On the Transfer of Energy in the Electromagnetic Field." The paper was published in *Philosophical Transactions of the Royal*

**FIGURE 1.42**
John Henry Poynting developed a method of calculating the energy in an electromagnetic wave.

*Society of London.* Poynting surmised that as the plane waves traveled they carried energy with them and that this energy per unit area per time (which we learned previously is power per unit area and we call it irradiance) can be calculated by what is now called the *Poynting vector.*

$$\bar{S} = \frac{1}{\mu_0}\left(\bar{E} \times \bar{B}\right). \tag{1.71}$$

For our *monochromatic plane waves* as given in Equations 1.64 and 1.65 propagating in the z direction and using Equation 1.69 the Poynting vector becomes

$$\bar{S} = \frac{1}{\mu_0}\left(E_0\hat{p}_E e^{j(kz-\omega t)} \times B_0\hat{p}_B e^{j(kz-\omega t)}\right)$$

$$= \frac{1}{\mu_0}E_0\frac{1}{c}E_0 c e^{2j(kz-\omega t)}\left(\hat{p}_E \times \hat{p}_B\right) \tag{1.72}$$

$$= c\varepsilon_0 E_0^2 e^{2j(kz-\omega t)}\hat{z}.$$

Since the electric field and magnetic field are orthogonal of each other and the direction of propagation, the Poynting vector must be incident on the x-y plane in the z direction as would be expected. Using Euler's identity and realizing that the real part of the identity is still a valid solution for Maxwell's equations then Equation 1.72 can be rewritten as

$$\bar{S} = c\varepsilon_0 E_0^2 \left( \cos\left(2(kz - \omega t)\right) + j\sin\left(2(kz - \omega t)\right) \right) \hat{z}$$

$$= c\varepsilon_0 E_0^2 \, \text{Real}\left\{ \cos^2(kz - \omega t) + j\sin^2(kz - \omega t) \right\} \qquad (1.73)$$

$$= c\varepsilon_0 E_0^2 \cos^2(kz - \omega t).$$

Note here that the values found from Equations 1.72 and 1.73 are equally valid solutions for the Poynting vector of the oscillating wave. This is an instantaneous value for the power per unit area, but what we really want to measure is the average power per unit area and we only need the average value of the sinusoidal oscillations represented by the exponential function or the cosine-squared function, respectively (which is 1/2). The actual irradiance of the plane wave incident onto the *x-y* plane is therefore

$$I = \langle S \rangle = \frac{|S|}{2} = \frac{1}{2} c\varepsilon_0 E_0^2. \qquad (1.74)$$

Now we have a complete wave theory that shows us how to determine light propagation through apertures and how to determine the irradiance due to the electromagnetic phenomena that Maxwell used to describe it. One thing to note here is while we have loosely mentioned color and therefore wavelength and frequency of the light, we haven't really pointed out what that means in terms of measureable quantities. Maxwell has shown us that light must be an electromagnetic wave with a wavelength and frequency. What wavelength and what frequency? Figure 1.43 shows a diagram of the

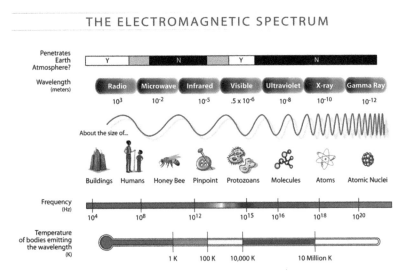

**FIGURE 1.43**
The electromagnetic spectrum. Credit NASA.

electromagnetic spectrum, as we know it today. Note that the wavelength of light is very small.

### 1.2.3.5 The Aether or Not?

Maxwell's equations made a bold impact on the scientific community in many ways. They unified electricity, magnetism, and light. But they also opened up a conundrum that is still discussed to date with some lack of clarity in many cases. This conundrum of how could a wave propagate through space when there is no physical medium for it to propagate through has perplexed the scientific community for well over a century. All other waves known at the time required some type of mechanical medium to travel through. The general belief of the 1800s and even early 1900s was that space had to be filled with some sort of material medium for a wave to propagate in.

This led to a very famous experiment known as the *Michelson-Morely experiment*. Since the belief was that there was a mechanical material medium filling space then as the earth moved through space and rotated about its axis, a "wind" from this material "aether" would be formed as it flowed by and around the planet. So, the American physicist Albert Michelson and American chemist Edward Morely devised an experiment in 1887 in which a very sensitive interferometer was used to attempt at detecting this "aether wind." The experiment never detected a mechanical medium and has been repeated over and over. The scientific conclusion was in and the "aether" must not exist yet somehow light waves didn't need it to propagate.

As far back as Plato and Aristotle, the great ancient philosophers theorized that the universe was filled with something, which they called the aether. Plato thought it was the "fifth element" and it was the "most translucent kind" of material. Aristotle had thought of it as some crystalline-like entity that he claimed was the "first element" rather than the fifth. He also noted that it was unlike the other elements and moved circularly and showed none of the characteristics of the other four elements. He thought it to be neither hot, cold, wet, nor dry. Aristotle's "first element" became known as the aether. In the 3rd century, Plotinus described the aether as though it penetrated all things and was nonmaterial. In the 1600s, the English polymath Robert Fludd also described the aether much in the same way as Plotinus as being prevalent throughout the cosmos and penetrating all matter while being a nonmaterial entity which was more "subtle than light" as he put it.

So, either the scientists of the 1800s (and perhaps even today) misunderstood what the properties of the aether must be or there isn't one and electromagnetic waves don't need a medium to propagate through. That said, the smart philosopher might say that space is the medium, so space is this so-called aether. However, space does not have the properties of a mechanical medium. It does on the other hand have properties of quantum physics, but we are getting ahead of ourselves. Suffice it to say that all experiments to date show that light indeed propagates through space whether it is a material medium or not!

### 1.2.4  1899–1930 A.D.

#### *1.2.4.1 Max Karl Ernst Ludwig Planck*

In 1899, the German theoretical physicist Max Planck (1858–1947), shown in Figure 1.44, began working on the physics of light and thermodynamics which led him to generate what has become known as Planck's black-body radiation law which describes the spectrum of light observed from a thermal electromagnetic source known as a black body. In order to make his radiation law match the observed phenomenon he had to make a bold adjustment to humanity's understanding of the universe and light in particularly. In 1901, he published the paper "On the Law of the Energy Distribution in the Normal Spectrum" in which he stated that electromagnetic energy can only be emitted in discrete or "quantized" amounts. This became known as *Planck's postulate* and is simply stated as

$$\varepsilon = nh\nu. \tag{1.75}$$

In Equation 1.75, $\varepsilon$ is energy (note the font is different from the electric field $E$), $n$ is the number of these quantized energy packets, $h = 6.626 \times 10^{-34}$ Js and is known as Planck's constant, and $\nu$ is the frequency of the electromagnetic wave. With Planck's postulate and Maxwell's equations, the calculation

**FIGURE 1.44**
German theoretical physicist Max Planck began working on the physics of light and thermodynamics which led him to generate what has become known as Planck's black-body radiation law.

of most phenomena regarding light was becoming possible. However, it was Planck's postulate that would soon cause the ugly particle versus wave argument to arise once again.

**Example 1.3: Calculating the Number of Quanta (Photons) in a Laser Beam**

A laser beam of orange-yellow light has a wavelength of 590 nm. The irradiance of the beam was measured to be 10 W/m² on a 1-cm-diameter detector plane. How many of Planck's quantized packets of light are incident on it per second?

First, we must determine the total amount of energy per second (which is power) being detected by finding the area of the detector, $A_d$, using the radius of the detector, $r$, as 0.5 cm or 0.005 m and

$$A_d = \pi r^2 = \pi (0.005\,\text{m})^2 = 7.85 \times 10^{-5}\,\text{m}^2. \tag{1.76}$$

The power incident on the detector is therefore

$$P = IA_d = \left(10\,\text{W/m}^2\right)7.85 \times 10^{-5}\,\text{m}^2 = 7.85 \times 10^{-4}\,\text{W}. \tag{1.77}$$

The energy incident on the detector in 1 s will be

$$\varepsilon = \frac{P}{t} = IA_d = \frac{7.85 \times 10^{-4}\,\text{W}}{1\,\text{s}} = 7.85 \times 10^{-4}\,\text{J}. \tag{1.78}$$

Using *Planck's postulate* (Equation 1.75), the value found for the energy in Equation 1.78, and $c = \lambda v$ then

$$\varepsilon = 7.85 \times 10^{-4}\,\text{J} = nhv = nh\frac{c}{\lambda}. \tag{1.79}$$

Solving for the number of quantized packets, $n$, results in

$$n = \frac{\varepsilon \lambda}{hc} = \left(7.85 \times 10^{-4}\,\text{J}\right)\frac{\lambda}{hc}$$

$$= \frac{\left(7.85 \times 10^{-4}\,\text{J}\right)\left(590 \times 10^{-9}\,\text{m}\right)}{\left(6.626 \times 10^{-34}\,\text{Js}\right)\left(2.997 \times 10^{8}\,\text{m/s}^2\right)} = 2.33 \times 10^{15}. \tag{1.80}$$

Wow! That's a lot of quantized packets per second just to make up a beam of only 785 μJ worth of energy! What does that suggest about the amount of energy in one of these packets?

## 1.2.4.2 Albert Einstein

In Example 1.3, we saw how using the tools to date that describe light will enable us to calculate some real tangible information about a laser beam of light. But we still aren't quite ready to understand what a laser is yet because we still haven't finished our discussion on what light is. Max Planck had postulated that the energy of the electromagnetic spectrum was discrete and could only exist in quantized amounts, but he didn't make the step in realizing that the electromagnetic field, and therefore light, itself might be discrete and what the implications of that conclusion might be.

In 1905, the German-born theoretical physicist Albert Einstein (1879–1955), shown in Figure 1.45, in his paper that translates as "Concerning an Heuristic Point of View Toward the Emission and Transformation of Light" explains the photoelectric effect (among other things) through the idea of Planck's quantized energy theory but takes it a step further in stating that,

> ...it seems to me that the observations regarding "black-body radiation," photoluminescence, production of cathode rays by ultraviolet light, and other groups of phenomena associated with production or conversion of light can be understood better if one assumes that the energy of light is discontinuously distributed in space. According to the assumption to be contemplated here, when a light ray is spreading from a point, the energy is not distributed continuously over ever-increasing spaces, but consists of a finite number of energy quanta that are localized in points in space, move without dividing, and can be absorbed or generated only as a whole.

In this paragraph, Einstein is telling us that he believed that light is made of discrete particles. His comments once again spark the particle versus wave debate but in a different way this time. The main difference is that Einstein also accepted right at the start of his paper that the wave theory of light

**FIGURE 1.45**
German-born theoretical physicist Albert Einstein described light as discrete "quanta localized in space."

...has worked well in the representation of purely optical phenomena and will probably never be replaced by another theory. It should be kept in mind, however, that the optical observations refer to time averages rather than instantaneous values. In spite of complete experimental confirmation of the theory as applied to diffraction, reflection, refraction, and dispersion, etc., it is still conceivable that the theory of light which operates with continuous spatial functions may lead to contradictions with experience when it is applied to the phenomena of emission and transformation of light.

What Einstein was saying in this paper was that he believed that from the macroscopic viewpoint over time lapses of the human observation capabilities (sight, touch, hear, smell, feel, etc.) that the wave theory should be accepted as a proper description of light. However, on the smaller, micro, and even nanoscopic scale and of very fast time increments a different theory is needed and hence his quantized particles. Einstein has given us the first acceptance of both ideas to describe one phenomenon in this paper. This is where the so-called "wave-particle duality" theory of light has its origin.

We should also note that Einstein follows along with Planck in calling these particles "quanta." In 1926, the American physical chemist Gilbert Newton Lewis (1875–1946) coined the term "photon" to describe Einstein's light particles. The idea that light was truly a wave-particle duality became the philosophy of the scientific community even though there was much heated debate over just what that truly meant.

What happened next in our understanding of light was many very brilliant scientists and engineers debated, tested, theorized, and experimented until quantum physics was created, tweaked, tweaked again, and tweaked some more (the tweaking never ends). The birth of quantum physics was in this time frame between Plank's work in 1901 and 1930. At this time, most of the scientists focused on light's interaction with matter and in so they developed a means to describe atoms and molecules in much the same way as they were beginning to describe light—as a wave-particle duality.

In 1913, the Danish physicist Niels Henrik David Bohr (1885–1962), shown in Figure 1.46, realized that light energy interacted with matter in these quantized amounts as proposed by Planck and Einstein. This allowed Bohr to develop the first decent model for the atom, which is all important in our development of how lasers work and will be discussed much further in Chapter 3. Bohr's model of the atom showed that electrons traveled about the nucleus in orbits, but these orbits were specific quantized energy states. When the right quanta of energy were incident on this atom then the electron would make a "quantum jump" (some say "quantum leap") to the next highest discrete energy level or "state." If for whatever reason the atom loses this energy (will be discussed in Chapter 3) then it only loses energy in discrete "quantized" amounts and the electron returns to the same lower-energy level orbit or "state." While Bohr's model for the atom was more

**FIGURE 1.46**
Danish physicist Niels Henrik David Bohr developed the basis for modern quantum physics and how light interacts with atoms.

focused on how light energy interacts with matter it does support the idea (and measurements) that light energy is indeed quantized as described by Planck's postulate in Equation 1.75.

Also at about the same time in a series of papers from 1911 to 1913 Max Planck calculated that the ground state of thermal material at absolute zero temperature was actually nonzero. This is often referred to as Planck's "second quantum theory." This was the first idea that there might be a ground-state energy filling the voids of the universe. Planck showed that this "zero point energy" was found as

$$\varepsilon_{zpe} = \frac{h\nu}{2}. \tag{1.81}$$

The idea became more important, especially as far as lasers are concerned, than the physicists of the time realized. Einstein even initially wrote a paper about it and then retracted it from publication because he wasn't sure he believed in the idea.

The early 1900s was a flurry of activity in the growth of modern theory of light quanta and their interactions, but there were many questions yet to be answered. The biggest question arising from Bohr's model (at least in my mind) is how does this quantized wave-particle duality thing now called a photon actually interact with an electron? Einstein had given the first real explanation of a process where his light quanta and electrons were connected in describing what happens when light is incident on a metal photoelectrode in the 1905 paper above. He showed that the energy of the electron that is emitted from the photoelectrode $\varepsilon$, is the Planck's quanta energy, $h\upsilon$, minus some work function of the material, $\phi$, or

$$\varepsilon = h\upsilon - \phi. \tag{1.82}$$

Even though Einstein's insight on the photoelectric effect shown in Equation 1.82 would be partly the reason for his winning a Nobel Prize in 1921, it only hinted at the underlying interactions of photons and matter that would be needed to move closer to what would become our modern theory of light. We should note here that Equation 1.82 is what allows laser scientists and engineers to implement photodetectors such as photomultiplier tubes and what would lead us to modern-day digital cameras and focal plane arrays.

In 1924, the French physicist Louis Victor Pierre Raymond de Broglie (1892–1987) expanded the hypothesis of the wave-particle duality from light quanta to all matter. In his doctoral dissertation, he surmised that all matter has wave properties and that any particle's momentum, $p$, describes its wave properties by

$$p = \frac{h}{\lambda}. \tag{1.83}$$

Equation 1.83 is *de Broglie's formula* and it describes all matter particles with momentum as having a wave property with a wavelength proportional to Planck's constant. What he actually hypothesized was that the electron in Bohr's orbit could actually be considered to be like a violin string tied to itself about the nucleus of the atom. That string was vibrating or oscillating with the wavelength as described in Equation 1.83 and each electron orbit had an associated vibrational energy related to it. At first, de Broglie's doctoral committee had a hard time accepting his hypothesis. Einstein stepped in and claimed that the idea, "…may look crazy but it is really sound!" de Broglie was not only awarded his PhD, but a few years later in 1929 was awarded the Nobel Prize for his "crazy idea."

At this point in history, humanity was on the verge of realizing that everything in the universe was pretty much a wave-particle duality conundrum with all sorts of strange and wonderful consequences to go along with that concept. Einstein and Bohr had very famous debates over the implications and meaning of Bohr's developing atomic model and light quanta interaction. Bohr was developing the belief that certain objects (in quantum mechanics) have "complementary properties" (such as being a wave and a particle) which cannot be observed or measured at the same time. A key belief Bohr had was that it is not possible to regard "quantum objects" as having independent properties from whatever measuring instrument is being used to "see" such properties. In other words, the "quantum object" and the measuring instrument become "connected" or "superimposed" upon one another.

In 1925, the German theoretical physicist Werner Karl Heisenberg (1901–1976), shown in Figure 1.47, published what is known as his "breakthrough" paper "Quantum theoretical re-interpretation of kinematic and mechanical relations." In this paper, he suggested that it was time to "…discard all hope of observing…the position and period of the electron…" at the same time. This was a more concise description of (or addendum to) Bohr's *principle of complementarity*.

While Heisenberg didn't completely formulate it, yet this within the 2 years to follow became the now famous *Heisenberg uncertainty principle*

$$\Delta x \Delta p \geq \frac{h}{2} \tag{1.84}$$

where $\Delta x$ is the uncertainty in position $x$, and $\Delta p$ is the uncertainty in momentum, $p$. The more modern form is

$$\sigma_x \sigma_p \geq \frac{\hbar}{2}. \tag{1.85}$$

Where $\hbar = h/2\pi$, $\sigma_x$ is the standard deviation in the position and $\sigma_p$ is the standard deviation in the momentum. According to Heisenberg, the principle merely states that

> One can never know with perfect accuracy both of those two important factors which determine the movement of one of the smallest particles— its position and its velocity. It is impossible to determine accurately *both* the position and the direction and speed of a particle at the same instant.

**FIGURE 1.47**
German theoretical physicist Werner Karl Heisenberg showed that both position and momentum can't be known about an object precisely.

About the same time frame, in 1927, Bohr publicly lectured on his *principle of complementarity* in Italy and slightly later in Belgium. With the complementarity and uncertainty principles, de Broglie's formula, Planck's postulate and zero point energy, and Einstein's continued overarching inputs (like the photoelectric effect) the piece parts for a new more "modern theory" of the interactions of light quanta, matter waves, and the cosmos itself (and soon lasers) were almost in place.

It was the Austrian physicist Erwin Rudolf Josef Alexander Schrodinger (1887–1961), shown in Figure 1.48, who finally put it all together by publishing several papers back to back much like Einstein had done in 1905. Schrodinger's paper "Quantization as an Eigenvalue Problem" gave a derivation for the time-independent wave equation that enabled him to calculate correct quantized energy levels for the hydrogen atom. The equation has since become known as *Schrodinger's equation* and it has also been written to account for time-dependent systems as well. There was now, due to Schrodinger, a mathematical formalism to describe all of matter and energy in the form of probabilistic quantum wave equations. In other words, for the first time in physics there was now a single equation that could be written out for almost any system interaction of matter and energy. The time-dependent Schrodinger equation is

$$-\frac{\hbar^2}{2m}\frac{\partial^2 \Psi(x,t)}{\partial x^2} + V\Psi(x,t) = i\hbar\frac{\partial \Psi(x,t)}{\partial t}. \tag{1.86}$$

And the time-independent Schrodinger equation is

**FIGURE 1.48**
Austrian physicist Erwin Rudolf Josef Alexander Schrodinger developed the wave equation that describes any interaction of matter and energy.

$$-\frac{\hbar^2}{2m}\frac{\partial^2 \psi}{\partial x^2} + V\psi = \varepsilon\psi. \qquad (1.87)$$

In Equations 1.86 and 1.87, $m$ is the mass of whatever particle/object we are discussing, $V$ is the potential energy, and $\psi$ is the time-independent portion of the complete wavefunction, $\Psi(x,t)$, where

$$\Psi(x,t) = \psi(x)e^{-\frac{i\varepsilon t}{\hbar}}. \qquad (1.88)$$

We should also point out here that the left-hand side of Equation 1.87 is the quantum mechanics equivalent of the *Hamiltonian* in classical mechanics and can be represented as the *Hamiltonian operator*, $\hat{H}$,

$$\hat{H} = -\frac{\hbar^2}{2m}\frac{\partial^2}{\partial x^2} + V. \qquad (1.89)$$

Therefore, Equation 1.87 can be simply written as

$$\hat{H}\psi = \varepsilon\psi. \qquad (1.90)$$

One of the most impactful and exciting aspects to the Schrodinger equation is that the solution to the time-dependent one always, in every single case, is a linear combination of wavefunctions of the form in Equation 1.88. In other words,

$$\Psi(x,t) = \sum_{n=1}^{\infty} c_n \psi_n(x)e^{-i\varepsilon_n t/\hbar} \qquad (1.91)$$

where $c_n$ are complex constants used to fit to the boundary conditions of the situation and $n$ is the summation integer corresponding to the particular function in the combination.

So what is so exciting about Equation 1.91? Well, according to Schrodinger's equation, which can be used to describe any physical system including light and matter then the solution turns out to be a superposition of wavefunctions. Philosophically, Equation 1.91 suggests that all interactions of quantum objects (which all things in the universe pretty much are at the molecular scale and below it seems) can be described as one set of wavefunctions interacting with another set of wavefunctions!

### Example 1.4: The Infinite Square Well

Figure 1.49 shows the idea of the so-called "infinite square well" of potential energy. The walls of the well are infinitely high in potential

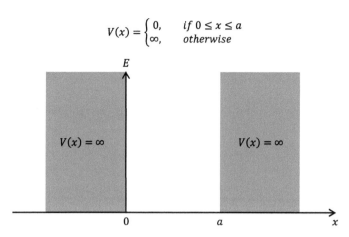

$$V(x) = \begin{cases} 0, & \text{if } 0 \le x \le a \\ \infty, & \text{otherwise} \end{cases}$$

**FIGURE 1.49**
The infinite square potential well.

energy value and the width of the well is some distance, *a*. Equation 1.92 describes the well mathematically as

$$V(x) = \begin{cases} 0 & \text{if } 0 \le x \le a \\ \infty & \text{otherwise} \end{cases}. \tag{1.92}$$

Imagine that some small particle is placed inside this well. That particle would be free to move about within the well; however, it so desired except at the boundaries of the walls since they are infinitely high. In other words, the particle can move about the well as it likes but it is trapped within it. So, what does Schrodinger's equation tell us about this particle?

First, we know that the particle can never be outside the well so the time-independent component of the wavefunction outside the well by definition of the well is

$$\psi(x) = 0 \tag{1.93}$$

Equation 1.93 means that the probability of finding the particle outside the well is zero. Inside the well, on the other hand, is a bit more complex. We know that the potential, *V(x)* is zero inside the well so the Schrodinger equation for the particle is

$$-\frac{\hbar^2}{2m}\frac{\partial^2 \psi}{\partial x^2} = \varepsilon \psi. \tag{1.94}$$

We can rewrite Equation 1.94 as

$$\frac{\partial^2 \psi}{\partial x^2} = -\frac{2m\varepsilon}{\hbar^2}\psi. \tag{1.95}$$

At this point, hopefully, we realize that Equation 1.95 is a differential equation that is in the form of the classical *simple harmonic oscillator*

$$\frac{\partial^2 \psi}{\partial x^2} = -k^2 \psi \tag{1.96}$$

If we set $k = \sqrt{2m\varepsilon}/\hbar$ and also note that $\varepsilon \geq 0$ then we can use the general solution to Equation 1.96 as

$$\psi(x) = A\sin(kx) + B\cos(kx). \tag{1.97}$$

$A$ and $B$ are arbitrary constants that can be found by knowing something about our boundary conditions. Since the walls of the well are infinitely high, we can use Equation 1.93 to see that

$$\psi(0) = \psi(a) = 0. \tag{1.98}$$

Therefore

$$\psi(0) = 0 = A\sin(0) + B\cos(0) = B. \tag{1.99}$$

Then since $B = 0$ Equation 1.97 becomes

$$\psi(x) = A\sin(kx). \tag{1.100}$$

Using the boundary condition for, $a$, we see

$$\psi(a) = 0 = A\sin(ka). \tag{1.101}$$

Equation 1.101 is a bit more tricky to understand. If $A = 0$ then the wavefunction is quite trivial and no fun at all. However, if we assume $A$ to be nonzero that tells us that

$$\sin(ka) = 0 \tag{1.102}$$

And that would mean that

$$ka = 0, \pm\pi, \pm 2\pi, \pm 3\pi, \dots. \tag{1.103}$$

All of the values for $ka$ in Equation 1.103 again give us the trivial solution of our wavefunction being equal to zero. The more useful scenario is when the wavefunction is nonzero when

$$k_n = \frac{n\pi}{a}, \qquad \text{where } n = 1, 2, 3, \dots \tag{1.104}$$

Equation 1.104 gives us a relation for the energy of the particle in the well found as

$$k_n = \frac{n\pi}{a} = \frac{\sqrt{2m\varepsilon}}{\hbar}. \tag{1.105}$$

Solving for $\varepsilon$ gives

$$\varepsilon_n = \frac{\hbar^2 k_n^2}{2m} = \frac{n^2\pi^2\hbar^2}{2ma^2}. \tag{1.106}$$

Equation 1.106 shows that a particle in this infinite well can only have discrete values of energy. In other words, the energy values allowed are quantized rather than continuous. Finally, in order to find the value of A we have to use a process called normalization. We integrate all of the total probable values of the wavefunction over all allowed values (in free space would be $-\infty$ to $\infty$) and set that equal to unity, but we know our particle only has nonzero values between 0 and $a$, therefore,

$$\int_0^a \psi^*\psi\,dx = \int_0^a \sin^2(kx)\,dx = |A|^2 \frac{a}{2} = 1. \tag{1.107}$$

So,

$$A = \sqrt{\frac{2}{a}}. \tag{1.108}$$

And finally, we have the time-independent Schrodinger equation for the particle in the infinite potential well that is found to be

$$\psi_n(x) = \sqrt{\frac{2}{a}}\sin\left(\frac{n\pi}{a}x\right). \tag{1.109}$$

Note that the importance of this example to understanding lasers will be much more obvious in the chapters to come. For now, take notice of the fact that the lowest nonzero energy state for the particle is at $n = 1$ with wavefunction and corresponding energy levels of

$$\psi_1(x) = \sqrt{\frac{2}{a}}\sin\left(\frac{\pi}{a}x\right) \tag{1.110}$$

$$\varepsilon_1 = \frac{\pi^2\hbar^2}{2ma^2}. \tag{1.111}$$

Equations 1.110 and 1.111 represent the "ground state" for the particle in the infinite potential well. The other energy levels are called "excited states." Figure 1.50 shows graphs of the ground state and first few excited states.

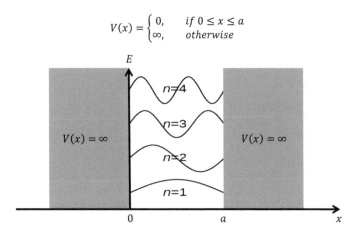

$$V(x) = \begin{cases} 0, & if\ 0 \le x \le a \\ \infty, & otherwise \end{cases}$$

**FIGURE 1.50**
The infinite square potential well ground state and first few excited states.

So why did we bother to look at this "particle" in this infinite square potential well? What we surmised from Example 1.4 is directly analogous to Bohr's model of the atom where the electrons orbit the nucleus and can only make quantum jumps from one state to the next and the natural state is the so-called "ground state." The other interesting point to note here is that when the "particle" jumps from one state to the next higher one, it must absorb energy of a quantized amount. For it to drop from an excited state to the next lower state it must then lose energy (emit it) in a quantized amount. Bohr actually showed that the quantized energy either emitted or absorbed was simply the energy difference between the energy levels of the two states involved and that it was due to a quanta of light at a given frequency. In other words,

$$\varepsilon_{n+1} - \varepsilon_n = h\nu. \tag{1.112}$$

At some point around 1925 in the development of quantum mechanics, Heisenberg wrote a paper to explain Bohr's model using his concepts of uncertainty (before the Schrodinger equation). But Heisenberg had yet to figure out all the math needed to solve the problem in depth. He sent his paper to the German physicist and mathematician Max Born (1882–1970), shown in Figure 1.51, who in turn involved his student Ernst Pascual Jordan (1902–1980), shown in Figure 1.52, in developing a matrix method to describe the energy and matter interaction of Bohr's model as well as the "harmonic oscillator." Using the matrix methodology, they analyzed many problems in quantum mechanics and developed a slightly different view than the Schrodinger wavefunction view that was coming soon. But it was Heisenberg, Born, and Pascual who generated the first algebraic formulation that would be used along with the wavefunction concept to bring forth modern quantum mechanics and therefore our modern theory of light.

**FIGURE 1.51**
German physicist and mathematician Max Born developed mathematical techniques that would enable modern quantum mechanics.

**FIGURE 1.52**
German physicist Ernst Pascual Jordan helped Max Born develop mathematical techniques that would enable modern quantum mechanics.

### 1.2.5 1930–2018 A.D.

We are almost to our present-day understanding of what we believe light is. The main pieces to the puzzle were mostly in place before 1930 or so, but there was much refinement to be done and a lot of philosophical interpretation and reinterpretation of what it all means. From 1930 to modern day, most scientists have tended to refer to light as "photons" and many do so without truly understanding what they are saying. It is often the case where the scientist or engineer will fall into the easy, and a bit sloppy, trap of saying light is particles when they need them to be and then they will turn around and use wave theory to solve a different problem never truly realizing, or caring about, what the wave-particle duality even means. And even worse, they will arrogantly state they understand what light is. Part of the reason for this lengthy first chapter is to aid the laser scientist and engineer in avoiding such academic fallacies. So, back to the story.

While Heisenberg, Born, and Pascual had developed a methodology and an algebraic language to use in describing energy's interaction with matter, and therefore light, it needed to be combined with de Broglie's ideas and the Schrodinger's method. It was the English theoretical physicist Paul Adrien Maurice Dirac (1902–1984), shown in Figure 1.53, who tied it all together in his first publication of *The Principles of Quantum Mechanics* in 1930. Most specifically to light and the description of lasers was his unique approach to solving the Schrodinger equation for the *simple harmonic oscillator*. The importance of knowing how to deal with a simple harmonic oscillator is that practically any oscillatory motion can be approximated as if it is simple harmonic motion within certain bounds (they must have small amplitudes).

The potential of the simple harmonic oscillator is parabolic and is found as

$$V(x) = \frac{1}{2}m\omega^2 x^2. \tag{1.113}$$

Where $\omega$ is the angular frequency found as

$$\omega = 2\pi v. \tag{1.114}$$

**FIGURE 1.53**
English theoretical physicist Paul Adrien Maurice Dirac created the first cohesive quantum theory in his first publication of *The Principles of Quantum Mechanics* in 1930.

Using Equation 1.113 the Schrodinger equation for the simple harmonic oscillator is written as

$$-\frac{\hbar^2}{2m}\frac{\partial^2\psi}{\partial x^2}+\frac{1}{2}m\omega^2x^2\psi=\varepsilon\psi. \tag{1.115}$$

There are two approaches to solving Equation 1.115 discussed in most texts on the subject. One approach requires a power expansion method for differential equations, which we will not go into here. The other is extremely clever and simple to apply. The origin of this second approach often called the "algebraic approach" is actually difficult to track down although most texts give Dirac credit for its invention. In fact, it does appear in his book and is one of the earliest known publications of the technique. The method is implemented by using *ladder operators*, which are sometimes called *creation-annihilation operators*. Again, they do appear in Dirac's book and he applies them ingeniously.

To start with, we must rewrite Equation 1.115 as

$$\left[\frac{1}{2m}\left(\frac{\hbar}{j}\frac{\partial}{\partial x}\right)^2+\frac{1}{2m}(m\omega x)^2\right]\psi=\varepsilon\psi. \tag{1.116}$$

Dirac realized that the term in brackets in Equation 1.116 can be written as the multiplication of two *factors* in the following form

$$u^2+v^2=(u-jv)(u+jv). \tag{1.117}$$

But what we have to realize in the case of the Schrodinger equation in Equation 1.116 that $u$ and $v$ are algebraic *operators* which by rule do not generally commute. In other words, usually $uv \neq uv$ although it does happen in some cases where they are equal by happenstance and the trivial case where they are both zero. Using Equations 1.116 and 1.117, we can develop our ladder operators. The *raising* or *creating* operator is

$$a\equiv\frac{1}{\sqrt{2m}}\left(\frac{\hbar}{j}\frac{\partial}{\partial x}+jm\omega x\right) \tag{1.118}$$

The *lowering* or *annihilation* operator is

$$a^\dagger\equiv\frac{1}{\sqrt{2m}}\left(\frac{\hbar}{j}\frac{\partial}{\partial x}-jm\omega x\right). \tag{1.119}$$

The "dagger" superscript in $a^\dagger$ denotes the fact that "a-dagger" is the adjoint of $a$. Also note that some texts will use a plus or minus sign rather than the dagger. Dirac actually used a bar over the letter to symbolize the lowering operator in his book. The dagger is one of the more common nomenclatures used in modern textbooks and thus we will use it herein.

Using Equations 1.118 and 1.119, we first need to test the operators to make certain they are correct. In other words, will they work as shown in Equation 1.117? Let's find out by multiplying them and operating them on a test function, $f(x)$

$$a^\dagger a f(x) = \frac{1}{\sqrt{2m}}\left(\frac{\hbar}{j}\frac{d}{dx} - jm\omega x\right)\frac{1}{\sqrt{2m}}\left(\frac{\hbar}{j}\frac{d}{dx} + jm\omega x\right)f(x). \qquad (1.120)$$

Start by performing the operators from right to left in Equation 1.120, so,

$$a^\dagger a f(x) = \frac{1}{2m}\left(\frac{\hbar}{j}\frac{d}{dx} - jm\omega x\right)\left(\frac{\hbar}{j}\frac{df}{dx} + jm\omega xf\right). \qquad (1.121)$$

Perform the second operator next

$$a^\dagger a f(x) = \frac{1}{2m}\left[\hbar^2\frac{d^2 f}{dx^2} + \hbar m\omega\frac{d}{dx}(xf) - \hbar m\omega x\frac{df}{dx} + (m\omega x)^2 f\right]. \qquad (1.122)$$

Finally, we see the result to be

$$a^\dagger a f(x) = \frac{1}{2m}\left[\left(\frac{\hbar}{j}\frac{d}{dx}\right)^2 + (m\omega x)^2 + \hbar m\omega\right]f(x). \qquad (1.123)$$

Now the really interesting bit of this check on our operators is that there is an extra term. Let's drop the test function and simplify it back into an operator form. The result is

$$a^\dagger a = \frac{1}{2m}\left[\left(\frac{\hbar}{j}\frac{d}{dx}\right)^2 + (m\omega x)^2\right] + \frac{1}{2}\hbar\omega. \qquad (1.124)$$

There is an extra $\frac{1}{2}\hbar\omega$ left over in Equation 1.124. Interestingly enough, if we reverse the order of the ladder operators we get

$$aa^\dagger = \frac{1}{2m}\left[\left(\frac{\hbar}{j}\frac{d}{dx}\right)^2 + (m\omega x)^2\right] - \frac{1}{2}\hbar\omega. \qquad (1.125)$$

Equations 1.124 and 1.125 show us that we can rewrite the Schrodinger equation if we just move the extra term to the other side. The Schrodinger equation from Equation 1.124 becomes

$$\left( a^\dagger a - \frac{1}{2}\hbar\omega \right)\psi = \varepsilon\psi. \tag{1.126}$$

Or from Equation 1.125 it is

$$\left( aa^\dagger + \frac{1}{2}\hbar\omega \right)\psi = \varepsilon\psi. \tag{1.127}$$

Here is the really ingenious part that Dirac's trick enables. The ladder operators can be implemented on the wavefunction in 1.126 or 1.127 and the Schrodinger equation stays the same with an added or subtracted quantized energy amount of $\hbar\omega$! In other words,

$$\left( a^\dagger a - \frac{1}{2}\hbar\omega \right)a^\dagger\psi = (\varepsilon - \hbar\omega)a^\dagger\psi. \tag{1.128}$$

And

$$\left( aa^\dagger + \frac{1}{2}\hbar\omega \right)a^\dagger\psi = (\varepsilon + \hbar\omega)a\psi. \tag{1.129}$$

At this point, I'm certain you are asking yourself all sorts of questions. What does this chapter all mean and why is it important to know as far as lasers are concerned? Equations 1.128 and 1.129 show us that a quantum oscillator, which might be an electron in an orbit about its nucleus, for example, can be described with the ladder operator and the wavefunction and that energy, a.k.a. light, will be emitted in a quantized amount if the oscillator is lowered a state. And likewise, energy, a.k.a. light, will be absorbed into the oscillator if it is raised by a state. We will discuss this much more in Chapter 3, but what we now have is a complete method of describing where light comes from (and sometimes where it goes) using the Schrodinger equation and the ladder operators.

### 1.2.5.1 Entanglement

From Dirac's publications in the 1930s to present day, our tools and techniques to describe light were mostly set. But our interpretation as to what these tools were telling us continued to change. Quantum mechanics gave us a probabilistic description of where light might be once it interacts with matter. In some instances, the mathematical tools suggest that the light might actually even be in two different states at once. This is the strange concept of *quantum entanglement*. The laser engineer might not run into a need for studying and understanding entanglement but the concept is becoming more and more developed and soon to be applied to technologies. So, any discussion on light and lasers would be incomplete without it.

In 1935, Einstein, along with Boris Podolsky and Nathan Rosen, pointed out that if quantum mechanics was correct then in some situations where two quantum particles are created within an event, the wavefunctions for those particles would become entangled and remain connected to each other. Their paper, now famously called the EPR paper, showed that these entangled particles can be very large distances away, but if one is interacted with the other, the other will instantaneously react as well. This instantaneous reaction is faster than the speed of light, which Einstein's theory of relativity doesn't allow. It should be noted here that in the EPR paper the "entanglement" was actually referred to as "spooky action at a distance." Schrodinger soon after wrote a paper defining what "entanglement" meant and it wasn't until 1957 when David Bohm actually connected "entanglement" to the EPR "spooky action at a distance."

Consider that some quantum event like a particle decay generates two photons traveling in opposite directions. Dirac developed tools to describe quantum particle spins and we now describe photons as having a spin of positive or negative one. The classical analog of spin might be a ball spinning about an axis. If the axis is pointing up and the ball spins to the right, it is a spin of 1. If the ball is spinning to the left then it is −1. But don't get too hung up on the classical analog because remember that light is not a particle or a wave or a ball with spin. It is some quantum object that has a property that we call spin. The property of spin is simply another detail about how certain photons react with other things in certain ways. The classical electromagnetic theory analog has to do with the propagation vector and which axis the electric field is normal too. Back to our story.

The two photons generated from the quantum event have opposite spin states. One of them will be in spin-up state (1) and one in spin-down state (−1). But due to the uncertainty of quantum mechanics we don't know which state and therefore we must write the wavefunctions of each photon, say $\psi_a$ and $\psi_b$ as being equally probable of being in either state. Therefore,

$$\psi_a = \frac{1}{\sqrt{2}}\left(\psi_1 + \psi_{-1}\right) \tag{1.130}$$

and

$$\psi_b = \frac{1}{\sqrt{2}}\left(\psi_{-1} + \psi_1\right). \tag{1.131}$$

The EPR paper pointed out that once the state of photon *a* (they actually discussed an electron positron pair but photons exhibit the same entanglement in this regard) was observed then photon *b* would immediately take the other state. Einstein didn't like this and thought that quantum mechanics was incomplete. His view at the time, along with many others incorrectly (even today), was that Equations 1.30 and 1.31 were merely probability equations telling us the likelihood of the photons being in one state or the other. Schrodinger pointed out a bit later (though he didn't believe it), and many scientists have argued this since, that the equations are actually not a probability description at all and in fact *are* the state of the photons. The photons truly are in both states at the same time!

Wait a minute? What? How can Equations 1.130 and 1.131 describe single photons each being in two states at the same time? This is because they are quantum things not particles or waves. They follow the rules of quantum mechanics. It is this point that is hardest for laser scientists and engineers to come to grips with typically because it means that a tangible thing such as a photon (which they have incorrectly thought of as particles) can intangibly be in two different states at the same time.

Let us obfuscate this even farther. Consider a single photon being incident on Young's two slits as described earlier in this chapter. There is a single photon incident on two slits. Common sense would tell us that the photon either goes through one slit or the other. The wavefunction for this photon can be represented much like Equation 1.31 and is

$$\psi = \frac{1}{\sqrt{2}}\left(\psi_{\text{slit }A} + \psi_{\text{slit }B}\right). \tag{1.132}$$

Again, a misguided laser scientist or engineer might suggest that Equation 1.132 shows the probability of the photon travelling through either slit A or slit B. Again that laser scientist or engineer would be incorrect. Equation 1.132 is the *actual* state of the photon. It actually *is* interacting with both slits at the same time. The outcome of the single photon interacting with the two slits is complicated and is still being debated today, but one interpretation is that there is a wavefunction describing the slits that in turn interacts with the wavefunction of the photon and the resultant reality is the superposition state of the two wavefunctions entangled into one. It is interesting to note here how in the modern era we say "wavefunctions" representing the information in light interacting with "wavefunctions" of objects which in turn creates what we see. By simply replacing the word "wavefunctions" with "simulacra," we are almost completely restating the *classic compromise* of the ancient era. Perhaps those ancient philosophers and great thinkers weren't so wrong after all!

Again, the ages old question of "what is light" is *still* perplexing and depending on the experiment, we might see different outcomes. In some experiments, a single photon detection occurs, but over time other single photons will build up an interference pattern even after the previous ones have hit the detector plane. How can this be? Dirac interpreted this interference conundrum in his 1930 book as:

> Each photon then interferes only with itself. Interference between two different photons never occurs.

It is unclear from this early statement in his book if Dirac only meant each photon in the specific two-slit experiment he was describing in the previous passage or if his statement was in general. Many modern textbooks treat his statement in the general. But what we know now is that this isn't necessarily the case. In 1964, the American theoretical physicist Richard Phillips Feynman (1918–1988), shown in Figure 1.54, published the now famous *Feynman Lectures on Physics*. In those lectures he addressed this conundrum as follows.

> One finds many books which say that two distinct light sources never interfere. This is not a statement of physics, but is merely a statement of the degree of sensitivity of the experiments at the time the book was written. What happens in a light source is that first one atom radiates, then another atom radiates, and so forth, and we have just seen that atoms radiate a train of waves only for about $10^{-8}$ sec; after $10^{-8}$ sec, some atom has probably taken over, then another atom takes over, and so on. So the phases can really only stay the same for about $10^{-8}$ sec. Therefore, if we average for very much more than $10^{-8}$ sec, we do not

**FIGURE 1.54**
American theoretical physicist Richard Phillips Feynman suggests that photons do interfere with each other.

see an interference from two different sources, because they cannot hold their phases steady for longer than $10^{-8}$ sec. With photocells, very high-speed detection is possible, and one can show that there is an interference which varies with time, up and down, in about $10^{-8}$ sec. But most detection equipment, of course, does not look at such fine intervals, and thus sees no interference. Certainly with the eye, which has a tenth-of-a-second averaging time, there is no chance whatever of seeing an interference between two different ordinary sources.

Recently it has become possible to make light sources which get around this effect by making all the atoms emit together in time. The device which does this is a very complicated thing, and has to be understood in a quantum-mechanical way. It is called a laser, and it is possible to produce from a laser a source in which the interference frequency, the time in which the phase is kept constant, is very much longer than $10^{-8}$ sec. It can be on the order of a hundredth, a tenth, or even one second, and so, with ordinary photocells, one can pick up the frequency between two different lasers. One can easily detect the pulsing of the beats between two laser sources. Soon, no doubt, someone will be able to demonstrate two sources shining on the wall, in which the beats are so slow that one can see the wall get bright and dark!

Feynman was indeed correct, and since 1964, this experiment has been accomplished and we realize that photons not only interfere with themselves but also with other photons if they interact in the right way. Dirac actually suggested something that would allow this by stating that "indistinguishable particles" will interfere and "distinguishable particles" will not. So perhaps he realized that his earlier statement was not in the general and that there could be and should be photon-to-photon interference if the photons looked exactly alike. In 1930, when Dirac wrote his book, however, the technology of the laser had yet to be invented and a source of such "indistinguishable particles" had yet to exist.

In 1963, the American physicist Leonard Mandel and colleague G. Magyar from the Imperial College of Science and Technology, London showed that light from two separate masers (as they were being called at the time) would in fact produce interference patterns if they were arranged in an appropriate manner. Feynman was actually suggesting that someday someone would do something that Magyar and Mandel had demonstrated the previous year in their paper published in *Nature* entitled, "Interference Fringes Produced by Superposition of Two Independent Maser Light Beams." It is unclear if Feynman knew of the publication or not and although the *Feynman Lectures on Physics* was published in 1964, it is likely that he had written that passage much earlier and it is possible that Magyar and Mandel's paper had yet to be published at the time of his writing that chapter of his book.

Feynman would go on to be integral in the development of Quantum Electrodynamics (sometimes called QED) and would be awarded a Nobel Prize in the process. QED is the more modern interpretation and mathematical construct used to consider how light and matter interact and is beyond the scope of this text. Suffice it to say that even Feynman recognized the strangeness of light and he appreciated the fact that we truly do not understand what it is. In his book *QED: The Strange Theory of Light and Matter* published in 1985, he explains

> It's rather interesting to note that electrons looked like particles at first, and their wavish character was later discovered. On the other hand, apart from Newton making a mistake and thinking that light was 'corpuscular,' light looked like waves at first, and its characteristics as a particle were discovered later. In fact, both objects behave somewhat like waves, and somewhat like particles. In order to save ourselves from inventing new words such as 'wavicles,' we have chosen to call these objects 'particles,' but we all know that they obey these rules for drawing and combining arrows [representing complex values of wavefunctions] that I have been explaining. It appears that all the 'particles' in Nature – quarks, gluons, neutrinos, and so forth (which will be discussed in the next lecture) – behave in this quantum mechanical way.

In this book, Feynman is telling us that light does have some macroscopic characteristics similar to waves and particles, but in reality they are things, objects perhaps is a better word, that follow the rules of quantum mechanics. This doesn't mean that quantum mechanics is complete and tells us everything there is to know about light. On the other hand, this "quantum object" description of light is the best one we have to date. Certainly, great thinkers of the future will look back on us and wonder how we could be so naive based on whatever their modern theory of light will be.

## 1.3 Chapter Summary

At the beginning of Chapter 1, we asked the question: What is light? In order to truly answer this question, we followed the answers given by many of the great thinkers throughout history. In Section 1.1, we discussed the classical description of light which took us from ancient times to the beginning of the mathematical era in the 1600s.

Section 1.2 is where we discussed the 1600s up until the modern era. This was the period of history where the modern mathematical understanding of how light interacts with things in the universe around us was being developed. During this era, the development of diffraction theory and interference was developed supporting the classical "wave theory." The electromagnetic theory of light was developed by Maxwell and then quantized by Planck and Einstein. A wave-particle duality interpretation for light was adopted following this for some time. With the advent of modern quantum mechanics and the development of the wavefunction description of light and matter interactions, the road to development of things like the laser was paved.

We are now mostly up to date on our understanding of light. For now, quantum theory is our best understanding. While there are some other even more confusing aspects to the quantum nature of light and matter interactions, there is no need to go into them here. For our purposes, we have learned the right answer to the question we asked at the start of this chapter. What is light?

Again, the answer is simply, light. What we do know about light is that it can be described as a quantum object or quantum particle, not to be confused with the classical idea of an object or particle, and that it follows the rules of quantum mechanics. In cases where we are averaging over long periods of time and looking at larger numbers of these quantum particles we can describe light as an electromagnetic wave that follows the vector calculus and the wave theory for diffraction. And in the cases of the small and discrete interactions with matter we can often times treat it as a particle that is quantized in energy composition. But in the end, it is best for us to realize the quantum nature of the beast and that light truly is just light!

## 1.4 Questions and Problems

1. What does the acronym LASER stand for?
2. What are the four elements of the *cosmogenic theory*?
3. What was significant about Greek philosopher Democritus of Abdera's "simulacra" when compared to modern quantum theory of light?
4. Who is credited as making the first "directed energy weapon"?
5. How were the contributions of Hero of Alexandria and Fermat important to our modern understanding of how light propagates along a path?
6. What experiment brought forth the end of Newton's corpuscular theory of light?
7. What is the Huygens–Fresnel principle?
8. Given a "beam director" aperture diameter of 1 m and a laser beam wavelength of 633 nm use the antenna designer's formula to estimate the far field of this optical system.
9. Consider a square aperture 1 cm by 1 cm. If a beam of 500 nm light is incident on it use Equation 1.25 to graph the far-field Fraunhoffer diffraction pattern.

10. Redo problem 9 but with a rectangular aperture of 1 cm by 2 cm instead.

11. Assuming that there is 1,550 W of sunlight per square meter incident on the surface of the earth and using 550 nm as the wavelength of the light (although we know it is spread across the visible spectrum from 400 to 700 nm) use Example 1.2 and Equation 1.39 to analyze Archimedes' "Death Ray." Assume the optical components are 80% reflective and that the ships of the Roman fleet were 2 km away. Assume the specific heat of spruce and pine is about $c \sim 2,300\,\text{J/kg }°\text{C}$. You will also need to recall the equation for heat is $Q = mc\Delta T$. Also assume the ignition temperature of wood is about 200°C.

12. If a "pinhole" can be represented as a delta function in space coordinate $x$, using Table 1.1 sketch the irradiance pattern expected to see in the far field of a beam passing through this pinhole.

13. From what was learned in Problem 12 what can be surmised about the light from a distant star once it reaches the earth?

14. If a star can be represented as a delta function or "point source" very far away, why do we see points in the sky at night rather than a uniform illumination from them as suggested by Problem 13?

15. Given:

$$\bar{E}(z,t) = E_0 \hat{x}_E e^{j(kz - \omega t)}$$

and

$$\bar{B}(z,t) = B_0 \hat{y}_B e^{j(kz - \omega t)}.$$

Assuming the wave is propagating in free space calculate the Poynting vector.

16. Calculate the irradiance in Problem 15.

17. A laser beam of blue light has a wavelength of 440 nm. The irradiance of the beam was measured to be 5 W/m² on a 1 mm diameter detector plane. How many photons of light are incident on it per second?

18. Why was Maxwell's addition of the displacement current so important?

19. Use Maxwell's equations to solve for the speed of light.

20. A cadmium photodetector has a work function of $6.52 \times 10^{-19}\,\text{J}$. What is the minimum wavelength of a photon required to eject a photoelectron from the detector? What happens if the energy of the photon is increased? What happens if two photons of the minimum required energy are incident on the detector? What does this suggest about irradiance versus wavelength?

21. What is the major significance of Equation 1.91?
22. Write down the Schrodinger equation using Hamiltonian operator.
23. Write down the Schrodinger equation using the ladder operators.
24. Write down the wavefunction for a photon incident on four slits.
25. (Challenge problem) Use what you have learned in this chapter to calculate the far field diffraction pattern of Young's two-slit experiment. Extrapolate from the diffraction theory to what might be seen if only one photon were incident on the slits. What if many photons are incident on the slits? What if some of the photons are distinguishable from one another?

# 2

## *What Is Amplification?*

As we discussed in Chapter 1 the word laser is an acronym that actually stands for Light Amplification by the Stimulated Emission of Radiation. We started by looking at the acronym itself as the simplest method to determine where to start in our study of the laser. The very first word in the acronym is "light." So, we discussed in great detail in Chapter 1 the answer to the question: "What is light?"

It makes sense that our next discussion should be the next major word in the acronym itself, which is "amplification." So what is amplification? In this chapter, we will discuss the details of what amplification is, what an amplifier is, and what are some of the figures of merit used to describe the process.

## 2.1 Amplifier Basics

### 2.1.1 Gain

Consider Figure 2.1, where a given signal, $f_{in}(t)$, is input into a system. The output signal, $f_{out}(t)$, is larger in amplitude and therefore is written as

$$f_{out}(t) = Gf_{in}(t) \qquad (2.1)$$

where $G$ is the amount of increase in amplitude of the input signal and is called the *gain* of the system. In other words, the *gain* of the system is determined by

$$G = \frac{f_{out}(t)}{f_{in}(t)}. \qquad (2.2)$$

This is an example of a simple *linear amplifier*. It should be noted that at first glance at Figure 2.1 the amplifier seems somewhat magical and amplifies the input signal and therefore delivering more output energy than input. This would indeed violate the law of conservation of energy. The diagram is most certainly oversimplified. Figure 2.2 shows a more accurate representation of a real-world amplifier where an external source of power must be implemented in order to amplify the input signal. It should be noted here that the gain of the system is not always a constant coefficient. The gain can

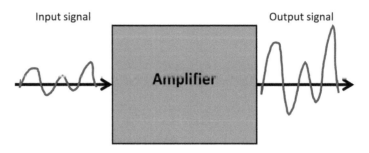

**FIGURE 2.1**
A basic amplifier increases the amplitude of the input signal.

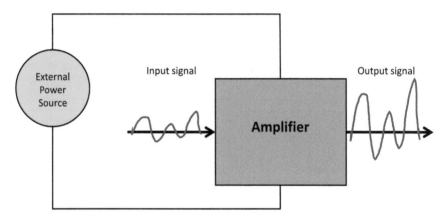

**FIGURE 2.2**
A basic amplifier requires an external power source.

be a constant, a linear function, and even a nonlinear function. Therefore, Equations 2.1 and 2.2 can be more precisely written as

$$f_{out}(t) = G(t) f_{in}(t) \qquad (2.3)$$

and

$$G(t) = \frac{f_{out}(t)}{f_{in}(t)}. \qquad (2.4)$$

Figure 2.3 shows a graph for $f_{in}(t) = \sin(t)$ and $f_{out}(t) = G(t)\sin(t)$. In this case the gain is constant with $G(t) = 10$. Figure 2.4 shows the same input function with $G(t) = t$. Figure 2.5 has a gain of $G(t) = f_{in}(t)$. Each of these Figures 2.3–2.5 illustrates how different functional forms of amplification can dramatically change the output signal of the amplifier system.

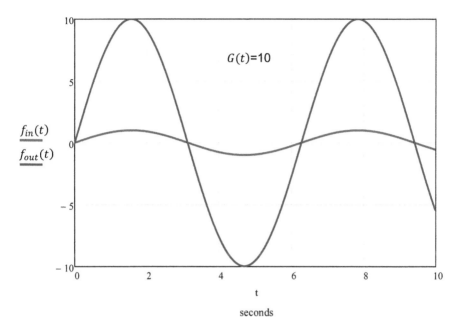

**FIGURE 2.3**
Output signal of a constant gain amplifier, $G = 10$, and a sinusoidal input.

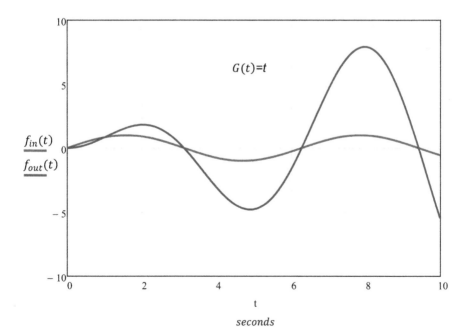

**FIGURE 2.4**
Output signal of an amplifier, $G(t) = t$, and a sinusoidal input.

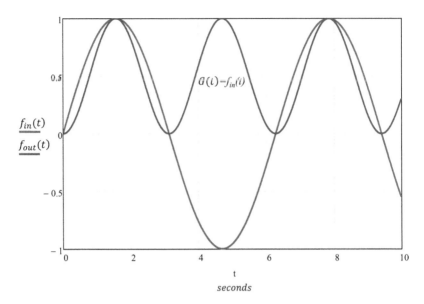

**FIGURE 2.5**
Output signal of an amplifier, $G(t) = f_{in}(t)$, and a sinusoidal input.

### 2.1.2 Saturation

Figure 2.6 shows again the same sinusoidal input as in Figures 2.3–2.5 but with $G(t) = e^{0.1t}$. Figure 2.7 shows a gain of $G(t) = t^2$. These two figures illustrate an interesting phenomenon in many amplifiers. At some point an amplifier can no longer increase the signal amplitude as it reaches the power available limit. This limit is known as *saturation*. For these graphs the limit was quite arbitrary and we could have arbitrarily increased the maximum of the y-axis range in order to see more of the waveform. In the real world, however, amplifiers have a maximum value of signal amplification and, likewise, output. Once the signal is amplified to a certain point it can no longer increase and therefore the tops (and bottoms) of the signal get cut off and the system will output the maximum or minimum value allowed. In this case, valuable waveform structure information can be lost. *Saturation* of amplifiers will be a very important phenomenon when discussing lasers as we will see in Chapter 4.

### 2.1.3 Frequency Response

Figure 2.8 shows a graph of the gain as a function of frequency, $G(\omega)$, for an arbitrary amplifier. This function is known as the *frequency response* of the amplifier. Once any amplifier has been constructed it is important to measure this parameter in order to know which frequencies can be amplified by the system and which ones cannot. Shown on the graph are the amplifier

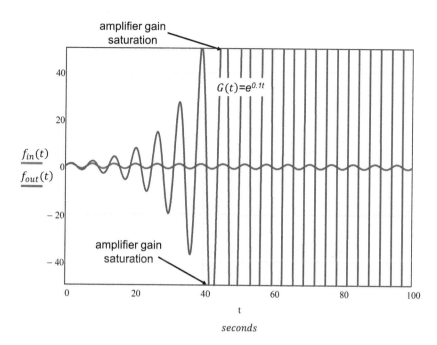

**FIGURE 2.6**
Output signal of an amplifier, $G(t) = e^{0.1t}$, and a sinusoidal input.

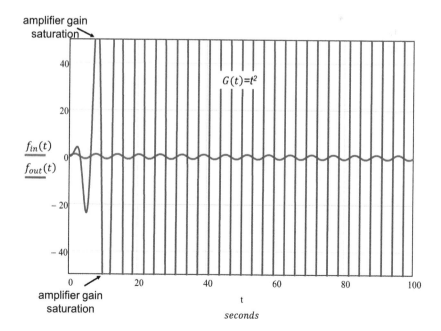

**FIGURE 2.7**
Output signal of an amplifier, $G(t) = t^2$, and a sinusoidal input.

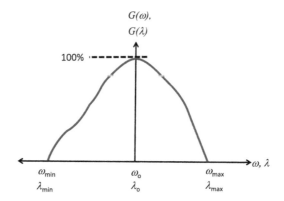

**FIGURE 2.8**
The frequency response curve of an amplifier shows the gain as a function of the frequency or wavelength.

*center frequency*, $\omega_o$, and the maximum and minimum *cut off frequencies* of the amplifier, $\omega_{max}$ and $\omega_{min}$, respectively.

Sometimes, especially with optical systems, the frequency response curve is given in terms of the wavelength, $\lambda$, rather than the frequency. This is simply a matter of convenience and actually, in some cases, nothing more than individual preference. More precisely, when using wavelength the function should be called the *wavelength response curve* of the amplifier, $G(\lambda)$.

## 2.1.4 Bandwidth

Figure 2.9 depicts another useful parameter when discussing amplifiers and that is the *bandwidth*. This feature is measured at the full width of the response curve at the half maximum amplitude value. The *bandwidth* at full width half maximum is $\Delta\omega_{fwhm}$.

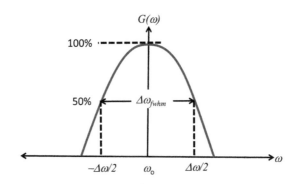

**FIGURE 2.9**
The bandwidth of an amplifier is measured at full width half maximum (FWHM) of the frequency response curve.

### 2.1.5 Noise

#### 2.1.5.1 Johnson–Nyquist Noise

An understanding of *noise* is extremely important for any laser scientist or engineer. One of the first types of noise the laser scientist might encounter is *Johnson–Nyquist noise*. It was first discovered by John B. Johnson in 1926 and was then explained by Harry Nyquist both of whom were working for Bell Labs at the time. *Johnson–Nyquist noise* is typically found in a system when random signal outputs occur due to thermal agitation of components within it. For this reason it is often referred to as *thermal noise.*

A good example of thermal noise is seen by heating a resistor within an electrical circuit. As the resistor is heated, it generates a random signal across most of the frequency spectrum and is therefore often used to simulate a uniform "white" noise spectrum.

When using light-detecting sensors such as cameras and photomultiplier tubes, warm detector planes can cause a signal to appear fuzzy or sometimes washed out. Often times, cooling the sensors will enable a much more precise detection by reducing the *thermal noise* of the detector. It should be noted here that this type of noise can occur whether power is being supplied to the detector or not. The detector will be at a particular temperature and therefore will be generating *thermal noise* based on that parameter. This type of noise is effectively a statement of Planck's law of black-body radiation which states that a system emits radiation based on its thermal situation with a spectrum defined by

$$I(\lambda) = \frac{2\pi hc^2}{\lambda^5} \frac{1}{e^{hc/\lambda kT} - 1}. \tag{2.5}$$

In Equation 2.5, $I(\lambda)$, is the *spectral irradiance* measured in power per unit area per unit wavelength, $h$ is Planck's constant, $c$ is the speed of light in meter per second, $\lambda$ is the wavelength in meters, $k$ is Boltzman's constant $(1.381 \times 10^{-23} \text{J/K})$, and $T$ is the temperature of the system measured in kelvin.

Figure 2.10 shows the *thermal noise* as calculated by Equation 2.5 for several temperatures. From the figure we can see that the hotter the system is the more *thermal noise* occurs as well as more noise power is shifted further into the visible. Figure 2.11 has a typical amplifier response curve overlaid on the *thermal noise* graphs. From this figure we must realize that a cooler system will be less noisy. This brings us to another parameter of amplifier systems known as the *signal-to-noise* ratio or *SNR*.

SNR is defined as

$$\text{SNR} = \frac{P_{\text{signal}}}{P_{\text{noise}}}. \tag{2.6}$$

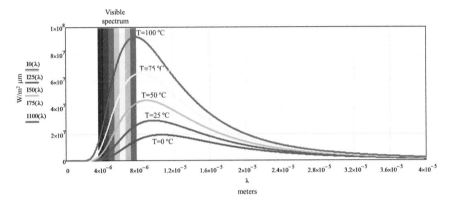

**FIGURE 2.10**
The spectrum of thermal noise for several temperatures as a function of wavelength.

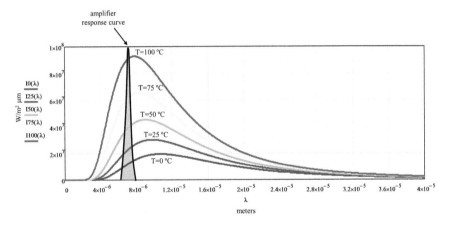

**FIGURE 2.11**
The spectrum of thermal noise for several temperatures as a function of wavelength compared with a typical amplifier response curve.

In Equation 2.6, $P_{signal}$ is the average power measured at the same point (in this case the same wavelength) as is the average power of the noise signal $P_{noise}$. Figure 2.12 shows values of SNR for the amplifier response curve's peak wavelength at temperatures 100°C, 75°C, 50°C, 25°C, and 0°C. Cooling the system has a dramatic impact on the SNR.

### 2.1.5.2 Schottky Noise

Another type of *noise* that will be of great importance to laser scientists and engineers is the so-called *shot noise*. This type of noise was first discovered in 1918 by Walter Schottky while studying current signals in vacuum tubes.

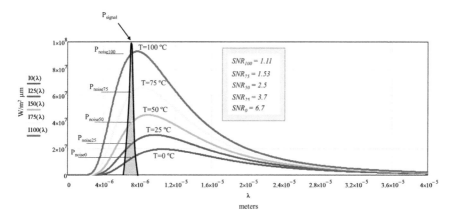

**FIGURE 2.12**
The SNR for a typical amplifier at various temperatures demonstrates the importance of controlling thermal noise.

Hence, it is sometimes (not as often as it should be) referred to as *Schottky noise*. It is also sometimes called *Poisson noise* because it can be modeled as a Poisson random measure or process. Schottky discovered that when a signal was generated by discrete phenomena like electrons or photons and if the measurement time of that signal was on a similar order of magnitude as the discrete particle arrival times then there might arise an error in the signal's measurement.

A good example of Schottky noise is by observing a tin roof during a strong rainstorm. Consider two measurement systems: (1) listening to the raindrops hit the roof and (2) measuring the flow of water out of a drain spout. We can hear the discrete raindrops hitting the roof and intuitively we understand that there are discrete particles hitting the roof. However, if our only measurement source were the drain spout then we would only see a continuous flow of water. When the signal is strong enough (meaning an abundance of discrete packets of energy, in this case raindrops) then our steady flow measurement will give us a good understanding of how much water is hitting the rooftop as a function of raindrops per second.

However, if we now consider this system during a very light sprinkling rain we will measure something entirely different. We will see only occasional spurts of water or even drops of water falling from the drain spout every so often although we can hear the rain hitting the roof with a steady "drop, drop, drop" sound.

Consider Figure 2.13 where we attach a flow meter to our drain spout in the above scenario. Assume that in order for the flow meter to measure the flow rate of rain pouring out of the drain spout it requires a 1-s integration time. From the figure it is shown that the flow meter will measure 11 raindrops per second hitting the roof plane for interval $t_0$ while instantaneously only three are. Also, by analyzing the five time intervals shown we see that

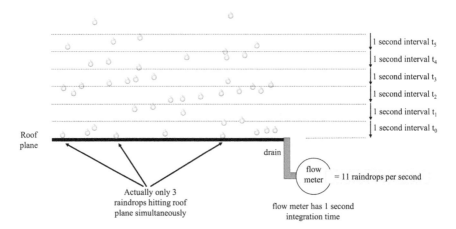

**FIGURE 2.13**
Detection of discrete particles or energy packets is accuracy limited by shot noise.

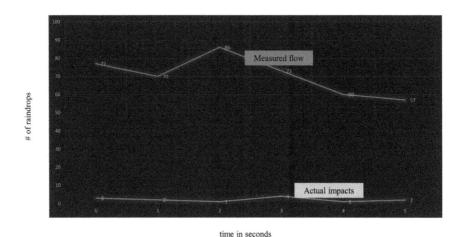

time in seconds

**FIGURE 2.14**
Measured flow rate and actual simultaneous raindrop impacts.

the flow meter would measure 5 drops for interval $t_1$, 12 for $t_2$, 10 for $t_3$, 7 for $t_4$, and 6 for $t_5$. Figure 2.14 shows a graph of the measured flow as well as the actual simultaneous drop impacts at each time interval start.

Now consider Figure 2.15 where the measurement intervals are started slightly later in time. The values from the flow meter will read 8 drops for interval $t_0$, 8 for $t_1$, 15 for $t_2$, 6 for $t_3$, 8 for $t_4$, and 3 for $t_5$. Figure 2.16 shows the graphs of the first scenario overlaid with that of the second. The numbers and graphs are significantly different based only on the fact that the detection time was slightly different.

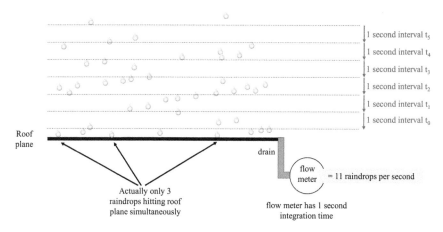

**FIGURE 2.15**
Detection of discrete particles or energy packets is accuracy dependent upon start time.

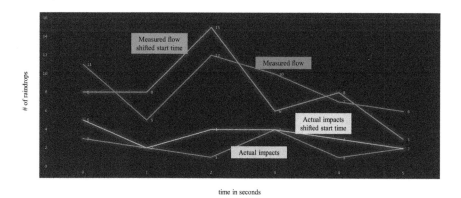

**FIGURE 2.16**
Measured flow rate and actual simultaneous raindrop impacts compared to measurements with a shifted start time.

Now consider Figure 2.17 where the number of raindrops falling per second is dramatically increased. There are many more raindrops hitting the roof and increasing the flow rate through the drain spout. Figure 2.18 shows the graphs of both sets of time intervals as with the previous scenarios in Figures 2.14 and 2.16. The graphs in Figure 2.18 show that with more raindrops falling per second, the data measured at different start times are more similar than the previous situation with fewer drops per second falling on the roof.

The discussion on *shot noise* thus far shows us that there is an uncertainty in detecting input signals based mainly on how and when it is measured. As Schottky discovered the uncertainty in the source signal, or *shot noise*, $\Delta\sigma$,

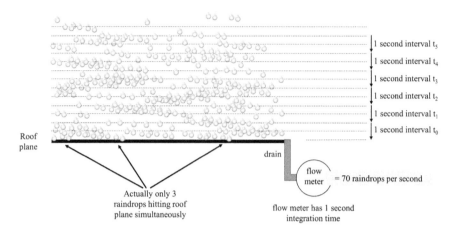

**FIGURE 2.17**
Detection of discrete particles or energy packets is accuracy dependent upon start time as well as the incident flux.

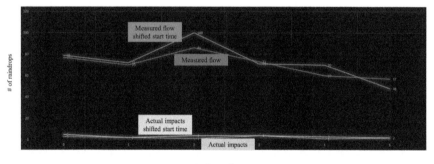

**FIGURE 2.18**
Measured flow rate and actual simultaneous raindrop impacts compared to measurements with a shifted start time.

is calculated as the square root of the total number of collected samples. In other words,

$$\Delta\sigma = \sqrt{N}. \tag{2.7}$$

The average number of raindrops hitting the roof per second, let's call the "raindrop flux", $I_0$. Therefore, the actual detected flux can be written as

$$I = I_0 + \Delta\sigma. \tag{2.8}$$

If we assumed that in our example above that we were receiving 11 raindrops per second for 1 second then the *shot noise* is 3.32 raindrops. In the second scenario where we were receiving roughly 70 raindrops per second the *shot noise* is 8.37 raindrops. This seems worse right? Actually, we have yet to consider the entire story. While Equation 2.7 tells us the uncertainty in our frequency measurement (in this case flow rate of raindrops on a roof) it doesn't tell us about signal strength versus the noise signal strength.

The SNR can be calculated by realizing that Equation 2.7 is the noise signal flux and Equation 2.8 is the total signal flux. Therefore,

$$\text{SNR} = \frac{I_0 + \Delta\sigma}{\Delta\sigma}. \tag{2.9}$$

Using Equation 2.9 we can compare the previous two scenarios. The SNR for the 11 raindrops per second scenario is 4.31 and for the 70 raindrops per second it is 9.36. Clearly the latter is a much better situation!

Laser scientists and engineers often find that laser amplifiers are limited by the *shot noise* of the system. It also becomes very important when detecting small numbers of photons in short increments of time with very fast laser pulses. One final note is that *shot noise* is also frequency independent and is therefore "white noise" and uniformly distributed about the frequency spectral response curve of an amplifier.

### 2.1.5.3 Noise Is Incoherently Considered

We have only gone into detail about two of the main types of noise that are inherent in laser amplifiers. There are many noise sources that the laser scientist or engineer might encounter. The noise sources are, as it turns out, incoherent and therefore can be considered independently in a square root of the mean sum of the squares. In other words, a system with $n$ noise sources will have a noise signal that is determined by

$$N = \sqrt{S_1^2 + S_2^2 + \cdots + S_n^2}. \tag{2.10}$$

## 2.2 Multiple Amplifiers

### 2.2.1 Amplifiers in Series

Figure 2.1 showed a basic amplifier and we learned some basics of amplifiers in Section 2.1. However, many systems become complex and have multiple amplifiers. Consider the system shown in Figure 2.19 with two amplifiers. These two amplifiers are in series with each other and the output of the first one is the input of the second. Using the amplifier equation

$$f_{\text{out}} = Gf_{\text{in}} \tag{2.11}$$

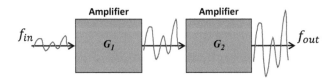

**FIGURE 2.19**
Two amplifiers in series.

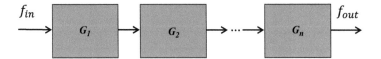

**FIGURE 2.20**
Multiple amplifiers in series.

we can determine the output of the series system as

$$f_{out} = G_2(G_1 f_{in}). \qquad (2.12)$$

If our system consists of $n$ amplifiers in series as shown in Figure 2.20 then Equation 2.12 can be written as

$$f_{out} = \prod_{1}^{n} G_n f_{in}. \qquad (2.13)$$

### 2.2.2 Amplifiers in Parallel

Another type of configuration that we will often see amplifiers in is parallel. Consider the system shown in Figure 2.21 with two amplifiers. These two amplifiers are in parallel with each other. In parallel, the input signal is split in half and sent into each one of the amplifiers (assuming they each have an equal impedance for electrical systems and/or we used a 50% beam splitter for an optical system). The signal is amplified by each with their respective gains and then recombined through addition. Using Equation 2.11 we can determine the output of the parallel system as

$$f_{out} = \left(G_1 \frac{f_{in}}{2}\right) + \left(G_2 \frac{f_{in}}{2}\right). \qquad (2.14)$$

If our system consists of $n$ amplifiers in parallel as shown in Figure 2.22 and we again assume $n$ equal split components of the input signal then Equation 2.12 can be written as

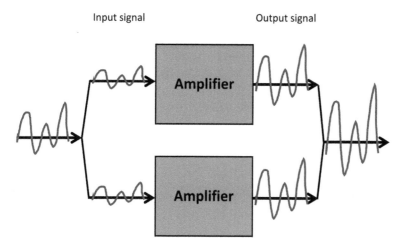

**FIGURE 2.21**
Two amplifiers in parallel.

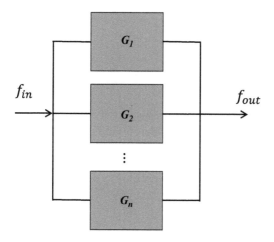

**FIGURE 2.22**
Multiple amplifiers in parallel.

$$f_{out} = \sum_{1}^{n} \frac{1}{n} G_n f_{in}. \tag{2.15}$$

It should be noted here that with parallel systems the larger the number of amplifiers, $n$, gets the smaller the input signal into each individual amplifier will be. At some point, the signal might be divided into too small a signal to do anything with! However, there are situations, such as in interferometers where parallel paths are needed.

### 2.2.3 Feedback Loops

Equations 2.13 and 2.15 will work well to describe noncomplex identical amplifiers in series and parallel. In most cases the system will be more complicated than that. If we have a system like that shown in Figure 2.23 where there is a single amplifier on the first pass through, but the output side is connected back through another amplifier with its output connected to the input signal creating a circuit loop. This loop is known as a *feeback loop*. The loop can be additive or subtractive in nature and therefore will be referred to as a positive or negative *feedback loop*. Figure 2.23 shows a positive *feedback loop* circuit.

Also note that the input signal, $x(t)$, gains, and output signal, $y(t)$, are shown as functions of time. For us to simplify analysis of these types of systems it will be useful to transform these functions into the Laplace domain using the Laplace transform given by

$$F(s) = \mathcal{L}\big[f(t)\big] = \int_0^\infty f(t)e^{-st}\,dt. \tag{2.16}$$

The inverse Laplace transform is

$$F(s) = \mathcal{L}^{-1}\big[F(s)\big] = \frac{1}{2\pi j}\int_{\sigma-j\infty}^{\sigma+j\infty} F(s)e^{st}\,ds. \tag{2.17}$$

Table 2.1 gives many common Laplace transforms and their inverses that will be useful in our amplifier systems analysis.

Now we will reconsider Figure 2.23 but in the Laplace domain. Figure 2.24 shows the system with transformed functions and gains. Note that a new function $E(s)$ was added as the input point to the first amplifier. We will start by writing an equation for this function as

$$E(s) = X(s) + G_2(s)Y(s). \tag{2.18}$$

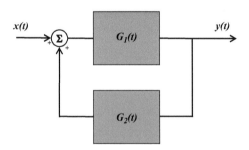

**FIGURE 2.23**
Amplifier in a positive feedback loop.

**TABLE 2.1**

Some Useful Laplace Transforms

$$\mathcal{L}\{1\} \rightarrow \frac{1}{s} \quad s > 0$$

$$\mathcal{L}\{e^{at}\} \rightarrow \frac{1}{s-a} \quad s > 0$$

$$\mathcal{L}\{t^n\} \rightarrow \frac{n!}{s^{n+1}} \quad s > 0$$

$$\mathcal{L}\{\sin(at)\} \rightarrow \frac{a}{s^2 + a^2} \quad s > 0$$

$$\mathcal{L}\{\cos(at)\} \rightarrow \frac{s}{s^2 + a^2} \quad s > 0$$

$$\mathcal{L}\{e^{at}\sin(bt)\} \rightarrow \frac{b}{(s-a)^2 + b^2} \quad s > a$$

$$\mathcal{L}\{e^{at}\cos(bt)\} \rightarrow \frac{s-a}{(s-a)^2 + b^2} \quad s > a$$

$$\mathcal{L}\{\sinh(at)\} \rightarrow \frac{a}{s^2 - a^2} \quad s > |a|$$

$$\mathcal{L}\{\cosh(at)\} \rightarrow \frac{s}{s^2 - a^2} \quad s > |a|$$

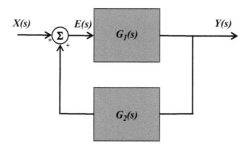

**FIGURE 2.24**
Amplifier in a positive feedback loop shown in the Laplace domain.

Solving for $X(s)$ results in

$$X(s) = E(s) - G_2(s)Y(s). \qquad (2.19)$$

We can also look at Figure 2.24 and realize that

$$Y(s) = E(s)G_1(s). \qquad (2.20)$$

Using Equations 2.19 and 2.20 to solve for the gain as in Equation 2.2, we find

$$G(s) = \frac{f_{\text{out}}}{f_{\text{in}}} = \frac{F_{\text{out}}}{F_{\text{in}}} = \frac{Y(s)}{X(s)}. \qquad (2.21a)$$

Also note that for the overall total gain of a complex system we will use the nomenclature $H(s)$, which is known as the transfer function of the system. Rewriting Equation 2.21a we find

$$H(s) = \frac{Y(s)}{X(s)} = \frac{E(s)G_1(s)}{E(s) - G_2(s)Y(s)}. \qquad (2.21b)$$

Simplifying Equation 2.21b and substituting Equation 2.20 for $Y(s)$ in the denominator gives us the equation for the transfer function of a positive feedback system

$$H(s) = \frac{G_1(s)}{1 - G_2(s)G_1(s)}. \qquad (2.22)$$

Likewise, the transfer function for a negative feedback system is

$$H(s) = \frac{G_1(s)}{1 + G_2(s)G_1(s)}. \qquad (2.23)$$

### Example 2.1: Amplifier with Feedback

Consider Figure 2.24 where the gains are both constant letting $G_1(t) = 1.25$ and $G_2(t) = 0.5$. Find the transfer function using Equation 2.22 and using Table 2.1 solve for the time domain representation of the output function.

First, we need to determine the gains in the Laplace domain using Table 2.1. Hence,

$$G_1(s) = \frac{1.25}{s} = 1.25\frac{1}{s}. \qquad (2.24)$$

And

$$G_2(s) = \frac{0.5}{s} = 0.5\frac{1}{s}. \qquad (2.25)$$

The transfer function is written as

$$H(s) = \frac{\dfrac{1.25}{s}}{1 - \dfrac{0.5}{s}\dfrac{1.25}{s}} = \frac{1.25s}{s^2 - 0.625}$$

$$(2.26)$$

$$= \frac{1.25s}{s^2 - 0.625} = 1.25\frac{s}{s^2 - \left(\sqrt{0.625}\right)^2}.$$

From Table 2.1 we find $h(t)$ is

$$h(t) = 1.25\cosh(0.625t)u(t) \tag{2.27}$$

where $u(t)$ is the unit step function. Therefore, we can write an equation for the output of the system as

$$y(t) = x(t)h(t) = 1.25\cosh(0.625t)u(t)x(t). \tag{2.28}$$

## 2.2.4 An Ensemble of Amplifiers

Consider the system shown in Figure 2.25 where there are multiple amplifiers in series as well as one in a feedback loop. This particular system is no more or less complicated than anything we've encountered thus far. However, in this case, we will add something new to the way in which our amplifiers in series function. For the moment we will only consider the series portion with an input function of $E(s)$ and an output function of $C(s)$. Typically, the gain of all the amplifier blocks could be represented as one function as described in Equation 2.13. But first, we will consider these amplifier blocks as individual copies of each other in a geometrically linear arrangement along a z-axis in spatial dimension. Assume they are literally connected exactly as shown in the figure and are indeed exact duplicates of each other so $G_1 = G_2 = \cdots = G_n$.

As the input signal propagates linearly along the z-axis, it is amplified more and more as it passes through each of the $n$ amplifiers. Therefore, an equation describing the overall amplification of a signal as it passes through the system from $E(s)$ to $C(s)$ can be written in terms of the variable $n$ as

$$\frac{dC(n)}{dn} = G(s)C(n). \tag{2.29}$$

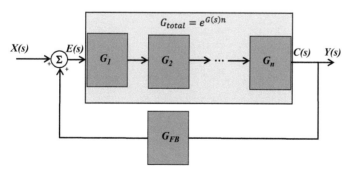

**FIGURE 2.25**
Multiple series amplifiers and a feedback loop positive gain system.

Rearranging the equation results in

$$\frac{dC(n)}{C(n)} = G(s)dn. \tag{2.30}$$

Integrating Equation 2.30 gives

$$\ln\left|\frac{C(n)}{C(0)}\right| = G(s)n \tag{2.31}$$

or

$$C(n) = C(0)e^{G(s)n}. \tag{2.32}$$

Equation 2.32 gives us a function for the overall gain of the series part of the system in Figure 2.25 which can be simply redrawn as a single amplifier component as shaded in gray in the Figure. With this simplified figure in mind the gain for the complete system can be found as

$$H(s) = \frac{e^{G(s)n}}{1 - G_{FB}(s)e^{G(s)n}}. \tag{2.33}$$

At first glance Equation 2.33 seems strange and not entirely useful. However, if the gain of each of the amplifiers in the series path is a constant and $n$ is just an integer constant and we assume the feedback gain is constant then the equation becomes

$$H(s) = \frac{e^{Gn}}{1 - G_{FB}e^{Gn}} = e^{Gn}\frac{s}{s^2 - \left(\sqrt{G_{FB}e^{Gn}}\right)^2}. \tag{2.34}$$

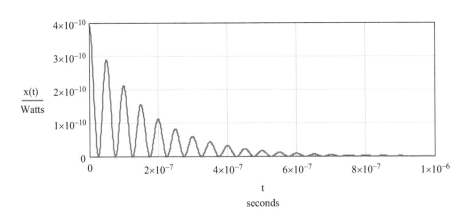

**FIGURE 2.26**
Arbitrary input light pulse for amplifier system shown in Figure 2.25.

Equation 2.34 is the transfer function for the complete system shown in Figure 2.25 and in the time domain the output is

$$c(t) = x(t)h(t) = e^{Gn}\cosh(G_{FB}e^{Gn}t)u(t)x(t). \tag{2.35}$$

Equation 2.35 is an important equation for laser scientists and engineers and we will see it again. For our purposes we will assume that we are always within the step function and therefore we can simplify Equation 2.35 as

$$y(t) = x(t)h(t) = e^{Gn}\cosh(G_{FB}e^{Gn}t)x(t). \tag{2.36}$$

Now let's consider an actual input function, $x(t)$ to be a pulse of light that can be described by

$$x(t) = N\frac{hc}{\lambda}e^{-2\pi ft}\cos(\alpha t) \tag{2.37}$$

where $N$ is the total number of photons in the pulse (assume 1 billion), $h$ is Plank's constant, $c$ is the speed of light in meter per second, $\lambda$ is the wavelength (assumed to be 500 nm), $f$ is the frequency of the pulse (assumed to be 1 MHz), and $\alpha$ is the angular frequency of the cosine function (assumed to be $2\pi \times 10^7$ rad/s).

At this point we also need to define the parameters for our amplifier. Let $n = 128$, $G_{FB} = 0.85$, and $G = 0.12$. Therefore, Equation 2.36 becomes

$$y(t) = e^{0.12(128)}\cosh(0.85e^{0.12(128)}t)\left(1\times10^9\frac{hc}{\lambda}e^{-2\pi(1\times10^6)t}\cos((2\pi\times10^7)t)\right). \tag{2.38}$$

Figure 2.26 is a graph of the input function, $x(t)$ and Figure 2.27 is a graph of the output function, $y(t)$. The input function amplitude is very small and

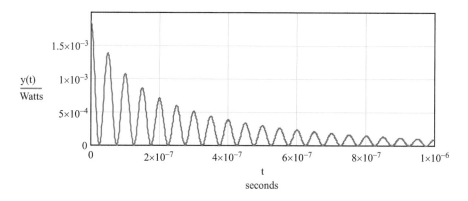

**FIGURE 2.27**
Output for amplifier system shown in Figure 2.25 with input shown in Figure 2.26.

indeed is amplified by many orders of magnitude. In fact, a nanowatt pulse is amplified to a milliwatt level pulse in this example.

By playing around with the numbers for the gains in Equation 2.38, we can quickly find regions where the system becomes unstable. In fact, if the gain, $G$, is only slightly higher the system will oscillate even without an input signal. In that case we would have a runaway amplifier and it would be of little use in the real world.

## 2.3 Chapter Summary

In our quest to develop an understanding of what a laser actually is, it makes sense that our second discussion following the one on light should be the next major word in the acronym, laser itself, which is "amplification." In this chapter, we discussed the details of what amplification is, what an amplifier is, and what are some of the figures of merit used to describe the process.

In Section 2.1, we covered many of the basic parameters and characteristics used to discuss and describe amplifiers including *gain, bandwidth,* and *SNR*. With these basic parameters we learned what amplifiers can do. Then in Section 2.2 we began applying that knowledge to systems of multiple amplifiers in *series, parallel,* and with *feedback loops*.

In Section 2.2, we showed how an ensemble of many amplifiers with feedback could be used to describe a real-world type of system. Making the use of the Laplace transform we learned how to analyze a system to determine the transfer function and therefore the output of a system based on a given input signal. While in this chapter it might not have been obvious just yet how some of the information is pertinent to the introduction of laser science and engineering, this will become much clearer in the following chapters.

## 2.4 Questions and Problems

1. What is *gain*?
2. Give an equation for a linear amplifier *gain*.
3. At some point an amplifier can no longer increase the signal amplitude as it reaches the power available limit. What is this limit known as?
4. What is the difference between the *frequency response curve* and the *wavelength response curve*?

5. What feature is measured at the full width of the response curve at the half maximum amplitude value?

6. What is another name for *thermal noise*?

7. Consider a system that is 100°C. If the output optical signal is very narrow in bandwidth (approximating a nanometer) with a center wavelength of 700 nm and has a peak irradiance at 100 W, use Equations 2.5 and 2.6 to estimate the SNR. Hint: graph Equation 2.5 to find the peak *thermal noise* at the center wavelength of the output.

8. What is Shot Noise also known as?

9. A photodetector collects and average of 777 photons in a microsecond interval. What is the *shot noise*?

10. In Question 9 what is the actual Irradiance that is incident on the detector?

11. Determine the SNR for Problem 9.

12. Using Figure 2.20 and the given information that the gain of each amplifier is 4, and $n = 12$, what is the gain of the system?

13. Using Figure 2.22 and the given information that the gain of each amplifier is 4, and $n = 12$, what is the gain of the system?

14. What is *feedback*?

15. Consider Figure 2.24 where the gains are both constant letting $G_1(t) = 1.5$ and $G_2(t) = 1.5$. Find the transfer function using Equation 2.22 and using Table 2.1 solve for the time domain representation of the output function.

16. Using Figure 2.25 and the given information that the gain of each amplifier is 4, $n = 12$, and the feedback gain is 0.5, what is the time domain output of the system?

# 3

---

# *What Is the Stimulated Emission of Radiation?*

As we discussed in Chapter 1, and will repeat here, the word laser is an acronym that actually stands for Light Amplification by the Stimulated Emission of Radiation. We started by looking at the acronym itself as the simplest method to determine where to start in our study of the laser. The very first word in the acronym is "light." So, we discussed in great detail in Chapter 1 the answer to the question: "What is light?"

It makes sense that our next discussion, as given in Chapter 2, was to ask about the next major word in the acronym itself which is "amplification." We discussed amplification and amplifiers in some detail. Now we can complete our analysis of the acronym by asking the question, "What is the *stimulated emission of radiation*?"

In this chapter, we will discuss the details of what stimulated emission of radiation is, why it is the major physical phenomenon at the heart of the laser, and how it is exploited in ways to create laser light beams. But before we go much further we need to make our nomenclature clear. What do we mean by "radiation"? In the context of the laser acronym, this does not mean radioactivity that one typically sees from an atomic nucleus decaying or being split or even fusing with another one.

What is meant by "radiation" in the context of our discussion is the exposure to or being immersed within an optical or light field from a classical perspective. From a more modern quantum mechanical perspective radiation means exposure to photons. One must keep in mind, again, that light is light, and the concept of the photon is much more complex than just assuming them to be particles. Therefore, being exposed to or having radiation emitted from a source means that the quantum wavefunction describing light is present and may or may not have interactions with the object or system in question.

Now that we understand what we mean by "radiation," we must ask ourselves "What is stimulated emission?"

## 3.1 The Bohr Model of the Atom

### 3.1.1 The Quantum Leap

Figure 3.1 shows a drawing of the atom as proposed by Niels Bohr in 1913. Known as the Bohr model of the atom, it was this first truly successful attempt to show how the structure of the atom interacts with energy that would eventually enable the development of the laser theory. It was Bohr's model that eventually led to the knowledge required to understand the phenomenology behind fluorescence and even lasers.

The model is quite simple in actuality. The atom consists of a nucleus with a positive charge of $Ze$ where $Z$ is an integer equal to the number of electrons orbiting the nucleus and $e$ is the net charge of the electron. Orbiting around the positively charged nucleus are negatively charged electrons in what Bohr called "stationary orbits" which has become known as just "orbits," "energy shells," or just "shells." The orbits or shells exist in subsequent layers or states as shown in Figure 3.1. These states are represented by the integer, $n$, which is known as the *principle quantum number* for the electron. The important piece of Bohr's model is the applied rule that electrons can only make integer jumps in energy from one of the orbits to the next and can never make partial state jumps. In other words, the electrons must make quantized jumps only; hence, the phrase "quantum leap" was born. The energy of these jumps is calculated by Equation 1.112. We will rewrite it here as

$$\Delta \varepsilon = \varepsilon_{n+1} - \varepsilon_n = h\nu. \tag{3.1}$$

where $\Delta \varepsilon$ is the quantized amount of energy emitted from the atom when the electron jumps to the next lower-energy shell. The atom emitting the photon

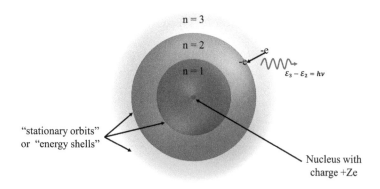

**FIGURE 3.1**
The Bohr model of the atom showing emission of a photon when an electron drops from one state to a lower one.

is known as *emission* and just what causes this to occur will be discussed in more detail in the next few sections of this chapter.

Figure 3.2 shows the same Bohr model of the atom but with the electron starting out on the $n = 2$ shell. A photon is absorbed by the atom adding the right amount of energy (a quantized amount) which in turn causes the electron to "leap" to the next higher shell. This process is known as *absorption*. The absorbed energy required to make the leap to the next higher-energy level is given by

$$\Delta \varepsilon = \varepsilon_n - \varepsilon_{n+1} = -h\nu. \tag{3.2}$$

The negative sign in front of the result simply shows that the energy was absorbed by the atom rather than emitted. The energy itself is still a positive quantity and is analogous to taking a bucket full of water from the ocean. The ocean would lose a bucket full of energy and the empty bucket would gain it.

Another point to be made about the atom in Figure 3.2 is that once the atom absorbs this photon and the electron moves to a higher state it is then said to be in an "excited state." In actuality, it is the electron itself that is in the higher than usual or excited state and when the reverse case happens and we see the events of Figure 3.1 occur and the electron drops back to the lower state by emitting a photon it is therefore de-excited. If all of the electrons of an atom are in the lowest states allowed by the quantum rules describing the atom then the electrons are said to be in the "ground state."

Figure 3.3 shows a helium atom which in the ground state has two electrons in the first shell. From high school chemistry we can recall that this is the $(1s)^2$ state where the superscript is the number of electrons in the subshell. Note that the electrons in the same orbit must also have opposite spins due to the *Pauli Exclusion Principle*. Upon the incidence of a photon the electron then

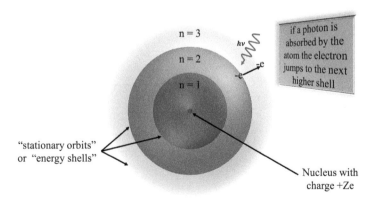

**FIGURE 3.2**
The Bohr model of the atom showing absorption of a photon when moving the electron from one state to a higher one.

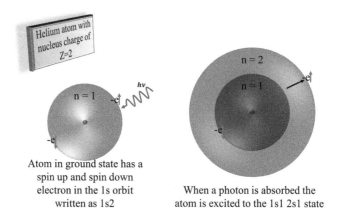

**FIGURE 3.3**
The Bohr model of the helium atom excited from the ground state to the first excited state by an incident photon.

leaps to the first excited state. The description of the electrons in the excited helium atom, therefore, becomes $(1s)^1 (2s)^1$. If the process reversed and the excited atom were de-excited by emitting the photon, the outer shell electron would drop back into the ground state. This process of excitation can be simplified by a two-dimensional graphic as shown in Figure 3.4. Versions of this graphic will come in handy for the laser scientist and engineer, as we will see more as this chapter continues.

Figure 3.5 shows two state models for the same helium atom. The state on the left side shows the electron leaping from the ground state to the excited state due to an incident photon. The right side of the figure shows

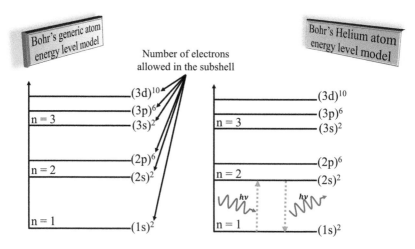

**FIGURE 3.4**
The Bohr model of the atom simplified into two-dimensional representation.

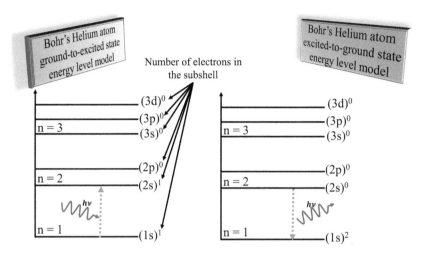

**FIGURE 3.5**
The Bohr model of the helium atom leaping from ground to excited states and back to ground again.

the electron in the excited state dropping back down to the ground state by emitting a photon.

### 3.1.2 Singlet, Doublet, and Triplet States

In Figures 3.1–3.5, we have shown the basic Bohr model of the atom where the electrons exist in shells and subshells about a positively charged nucleus. As the electrons become excited to outer shells and de-excited to lower shells we see this is due to either the emission or absorption of a photon of appropriate energy level. But depending on the shell and the number of electrons in the shells and subshells the type of excitation state might differ.

According to the *Pauli Exclusion Principle,* no two *fermions* (particles with ½ integer spins such as electrons) can occupy the same quantum state within the same quantum system at the same time. In other words, they cannot have the same quantum numbers describing the shell and spin they occupy or they would be the same electron. More easily stated, we can think of this principle as telling us that if electrons are in the same subshell then they must have opposite spins and the total number of those spins should equal to zero. For example, the helium atom has two electrons in the first shell. One electron is +½ spin and the other is −½ spin and therefore the sum of the spins equals zero. It is said that these electrons are in the *ground singlet state.* If one of the electrons gets excited to the next energy level it typically maintains its spin state and will, therefore, move to an *excited singlet state.* Figure 3.6 shows a depiction of a system in the ground singlet state being excited to a singlet state. Note that for such a system to be in the singlet state the sum of the spins in the system must still equal zero. The process for a

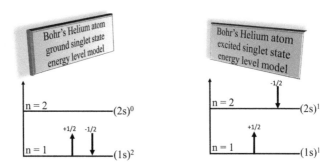

Sum of the ½ integer spins must equal zero

**FIGURE 3.6**
The singlet state has a zero net spin.

system absorbing a photon and being excited to the singlet state and then emitting a photon and dropping back down to the ground singlet state is called *fluorescence*.

Figure 3.7 shows a system in the *doublet state*. For a system to be in the doublet state the sum of the spins of the quantum system must equal ½. Figure 3.8 shows a system in the triplet state. The figure on the left side of Figure 3.8 shows the case of a spin flip within the same shell and on the right side shows an excitation to the next level. For the electron to drop back to the ground singlet state it should emit a photon. This process is called *phosphorescence*. It will be important later to note that fluorescence usually occurs many orders of magnitude faster than phosphorescence.

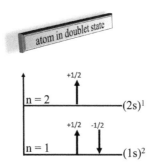

Sum of the ½ integer spins is equal to ½

**FIGURE 3.7**
The doublet state has a ½ net spin.

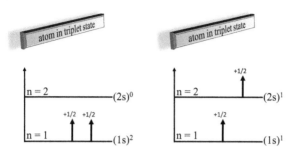

Sum of the ½ integer spins is equal to one

**FIGURE 3.8**
The triplet state has a net spin of one and can be a spin-flipped excitation in the same shell as on the left or a higher one as shown on the right.

## 3.2 Absorption, Stimulated Emission, and Spontaneous Emission

### 3.2.1 The Einstein Coefficients

In 1916, Albert Einstein developed the concept that light and matter (atoms and electrons) interact in three ways as shown in Figure 3.9. He developed the concept before quantum mechanics was complete using a thermodynamic argument. Consider that there are a total number, $N$, of atoms in a box of some sort. Now let's assume that there are $n_2$ in the excited singlet state and $n_1$ in the ground singlet state. Einstein surmised

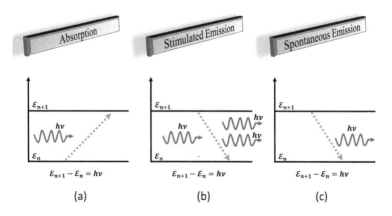

**FIGURE 3.9**
The three ways that light can interact with matter: (a) absorption, (b) stimulated emission, and (c) spontaneous emission.

that the decay rate for an excited atom to drop to the ground state could be represented as the related rate

$$\frac{dn_2}{dt} = -A_{21}n_2.$$ (3.3)

$A_{21}$ is the Einstein "A" coefficient for spontaneous emission. Likewise for atoms in the ground singlet state, $n_1$, to be excited to the first singlet state when exposed to a light field with spectral energy density, $\rho(v)$ is expressed as

$$\frac{dn_1}{dt} = -B_{12}n_1\rho(v).$$ (3.4)

$B_{12}$ is the Einstein "B" coefficient for the absorption rate of this particular system of atoms in a box. This coefficient can also be used to determine the rate at which the incident light field will "stimulate" emission by

$$\frac{dn_1}{dt} = B_{21}n_2\rho(v).$$ (3.5)

$B_{21}$ is the Einstein "B" coefficient for the stimulated emission of photons from level 2 to level 1 due to the incident light field. Realizing that there are a finite number of atoms, $N$, in our box, we can then write an equation for the states of the atoms when there is an equilibrium of atoms in each state as

$$B_{12}n_1\rho(v) = A_{21}n_2 + B_{21}n_2\rho(v).$$ (3.6)

The left-hand side of the equation represents the atoms in the ground state being excited through absorption while the right-hand side represents the atoms in the excited state decaying via spontaneous or stimulated emission.

Solving for the spectral energy density field in terms of the coefficients gives

$$\rho(v) = \frac{A_{21}n_2}{B_{12}n_1 - B_{21}n_2}.$$ (3.7)

Simplifying Equation 3.7 leads to

$$\rho(v) = \frac{\dfrac{A_{21}}{B_{12}}}{\dfrac{n_1}{n_2} - \dfrac{B_{21}}{B_{12}}}.$$ (3.8)

We know from thermodynamics and the *Maxwell–Boltzman distribution* that for a constant temperature, $T$, then

$$\frac{n_1}{n_2} = e^{\frac{hv}{kT}}$$

(3.9)

where $k = 1.38064903(51) \times 10^{-23}$J/K and is known as Boltzman's constant. Now we can rewrite Equation 3.8 as

$$\rho(v) = \frac{A_{21}}{B_{12}} \left[ \frac{1}{e^{\frac{hv}{kT}} - \frac{B_{21}}{B_{12}}} \right].$$

(3.10)

Also from thermodynamics, Planck's Radiation Law can be written as

$$\rho(v) = \frac{8\pi h v^3}{c^3} \left[ \frac{1}{e^{\frac{hv}{kT}} - 1} \right].$$

(3.11)

Comparing Equations 3.10 and 3.11, we can then write

$$\frac{B_{21}}{B_{12}} = 1$$

(3.12)

or

$$B_{21} = B_{12} = B$$

(3.13)

and

$$\frac{A_{21}}{B_{12}} = \frac{8\pi h v^3}{c^3}.$$

(3.14)

So, Equation 3.11 becomes

$$\rho(v) = \frac{A_{21}}{B} \left[ \frac{1}{e^{\frac{hv}{kT}} - 1} \right].$$

(3.15)

Einstein realized that Equation 3.15 tells us that there is a situation where an energy field density exists that is due to the absorption of incident energy and the spontaneous and stimulated emission of energy, which might be a sustainable phenomenon. He developed this derivation to explain the atomic spectral line emissions being measured and observed at the time. While this is the basis needed to move forward in defining how a laser works, it is unclear whether Einstein realized this at the time or not.

What Einstein did tell us through this derivation is that light will be absorbed by atoms and subsequently the atoms can emit light through the two processes of spontaneous or stimulated emission. He also did mention the fact that photons prefer to travel together in the same state. This is the idea that a photon of the right frequency incident upon an excited atom will stimulate the atom to emit a duplicate photon in the same frequency and phase as the original. This is the basics to the idea of *coherence*.

### 3.2.2 An Aside on Spontaneous Emission

Consider an atom with an electron in the outer shell being excited in a steady-state electric field. As the atom is excited each time, we will turn off the external field. One might think that there is no reason for that atom to fall back to the ground state. Unless there were another photon (an external electromagnetic field) incident on it, what triggers the de-excitation or the emission of a photon to occur spontaneously?

If we realize that this electron bouncing up and down from state to state is for all intents and purpose a simple oscillator we can make some interesting analyses from information we've already learned. Recall our discussion from Section 1.2.5 on the quantum wavefunction and the Schrodinger equation for the simple harmonic oscillator. Recall Equation 1.119 where the lowering ladder operator for going down the ladder is given and let us define a lowest rung of the ladder in a simple harmonic oscillator where

$$a^{\dagger}\psi_0 \equiv 0. \tag{3.16}$$

Using Equation 1.119, we see that

$$a^{\dagger}\psi_0 = 0 = \frac{1}{\sqrt{2m}}\left(\frac{\hbar}{j}\frac{\partial\psi_0}{\partial x} - jm\omega x\psi_0\right). \tag{3.17}$$

Simplifying Equation 3.17 gives

$$\frac{\partial\psi_0}{\partial x} = -\frac{m\omega}{\hbar}x\psi_0. \tag{3.18}$$

Rewriting Equation 3.19 and integrating

$$\int\frac{d\psi_0}{\psi_0} = -\frac{m\omega}{\hbar}\int x\,dx \tag{3.19}$$

gives the following result

$$\ln(\psi_0) = -\frac{m\omega}{2\hbar}x^2 + \text{constant}. \tag{3.20}$$

Solving Equation 3.20 for the wavefunction

$$\psi_0 = e^{-\frac{m\omega}{2\hbar}x^2} e^{\text{constant}}. \tag{3.21}$$

Let

$$A_0 = e^{\text{constant}}. \tag{3.22}$$

Finally, we see that the ground-state wavefunction for the system is

$$\psi_0 = A_0 e^{-\frac{m\omega}{2\hbar}x^2}. \tag{3.23}$$

Using Equations 3.16 and 1.127, we can solve for the energy level of the ground state by

$$\left(aa^+ + \frac{1}{2}\hbar\omega\right)\psi_0 = \varepsilon_0\psi_0 \tag{3.24}$$

$$a a^+\psi_0 + \frac{1}{2}\hbar\omega\psi_0 = \varepsilon_0\psi_0 \tag{3.25}$$

$$\frac{1}{2}\hbar\omega_0 = \varepsilon_0. \tag{3.26}$$

Equation 3.26 shows us that the ground-state energy, $\varepsilon_0$, is equal to a half quanta or half a photon energy. We know that photons do not exist as half photons, don't we? So what does this mean?

Without reading too much into Equation 3.26 what we can say is that the ground-state energy of the simple harmonic oscillator is nonzero. This turns out to be the same solution that Planck calculated for the ground state of a thermal material at absolute zero as shown in Equation 1.81 restated below

$$\varepsilon_{zpe} = \frac{h\nu}{2} = \frac{1}{2}\hbar\omega_0. \tag{3.27}$$

This "thermal material" is the vacuum of space for all intents and purposes. In other words, as discussed in Chapter 1 there is a vacuum field or "zero point energy" that might be considered the ground state of the universe that is filled with half quanta energy. The physical manifestation of this energy is often debated and beyond the topic of this discussion, but what is most germane to spontaneous emission is that our single atom with the excited electron is NEVER not in a nonzero electromagnetic field. It is this ground-state zero point energy that is always there and can and does "stimulate" emission of photons. What we learn from this is that there truly is no such thing as spontaneous emission.

In all honesty, Equation 3.3 should be rewritten with this knowledge as

$$\frac{dn_2}{dt} = -A_{21}n_2\rho_{zpe}(v). \tag{3.28}$$

While Equation 3.28 is merely "made up" to express a point about spontaneous emission, it is not typically implemented in any fashion. That said, what can be ascertained about the phenomenon comes from our discussion of *Heisenberg's uncertainty principle* in Chapter 1. Equation 1.84 was given in terms of position and momentum, but it can also be given in terms of energy and time as shown here

$$\Delta\varepsilon\Delta t \geq \frac{\hbar}{2}. \tag{3.29}$$

Equation 3.29 is often interpreted as though there is a minimum amount of time that exists where some of this half quanta ground-state zero point energy can "fluctuate" from its esoteric virtual realm into a measureable real photon in our universe. In other words

$$\Delta\varepsilon \geq \frac{\hbar}{2\Delta t} \geq \hbar\omega. \tag{3.30}$$

Where in this case $\hbar\omega$ is the energy of the real photon. It is believed that it is this photon that triggers or stimulates the emission process during spontaneous emission. Since this photon cannot exist for very long, it quickly vanishes back into the vacuum and therefore we only see one photon emitted during the process.

Again, this is the most common modern interpretation of what is going on. However, we have shown that the vacuum ground-state energy is nonzero in Equation 3.27. Through an interpretation of spontaneous emission as given in Equation 3.28, perhaps it is not necessary for the "virtual photon" to "pop" into reality at all. Perhaps the vacuum nonzero field is really all that is required to stimulate the spontaneous emission. This concept is still debated and open to interpretation. Hopefully, some clever laser scientist or engineer in the not too distant future will devise an experiment to determine a more precise understanding of the phenomenon.

### 3.2.3 Excited State Decay Rate

Consider once again our box full of $N$ atoms. Assume that $n_2$ of them are in the excited state. In Equation 3.3 (and perhaps more precisely in Equation 3.28), we showed that these excited atoms will decay from the excited state back to ground through spontaneous emission. The solution to Equation 3.3 can be written as

$$n_2(t) = n_2(0)e^{-A_{21}t}. \qquad (3.31)$$

Equation 3.31 is a common decay rate equation where the *radiative lifetime, τ,* of the decay is

$$A_{21} = \frac{1}{\tau}. \qquad (3.32)$$

And Equation 3.31 becomes

$$n_2(t) = n_2(0)e^{-\frac{t}{\tau}}. \qquad (3.33)$$

Again Equation 3.33 is a typical population decay model that, with no external energy input, decreases toward zero based on the *radiative lifetime, τ,* which can vary in value from nanoseconds to milliseconds depending on the type of radiative process and the electron transitions that occur. Singlet-state transitions are typically in the nanoseconds while triplet-state transitions can take three or more orders of magnitude longer to decay.

There are also nonradiative transitions where instead of emitting a photon to change energy levels a *phonon* is emitted. A *phonon* is a vibrational quanta of energy where one atom can transfer energy to another through physical means other than electromagnetic transfer. The classical concepts of heat and mechanical vibration transfer energy quantum mechanically through *phonons.* The energy of the *phonon* is quantized and can be determined by the spacing between atoms and the boundary conditions containing them. The quantized amount of phonon energy is not necessarily the same as the quantized amount of the photon energy but they are still quantized via Planck's constant as

$$\varepsilon_n = \left(\frac{1}{2} + n\right)\hbar\omega_{\text{phonon}} \qquad \text{for } n = 0, 1, 2, 3.... \qquad (3.34)$$

Interestingly enough, the phonon energy does come into play when discussing our $N$ atoms in a box. Figure 3.10 shows a two-level energy diagram of a singlet-state absorption and emission and a three-level energy diagram of a singlet absorption, nonradiative or phonon decay to a triplet state, and then

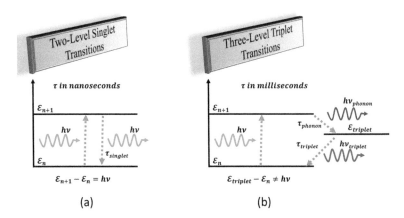

**FIGURE 3.10**
Two-level singlet transitions versus three-level triplet transitions: (a) two-level singlet radiative lifetime is typically nanoseconds and (b) three-level triplet transitions have longer radiative lifetimes.

a radiative decay with a triplet to ground-state photon emission. The lifetime of the singlet-state transition can be determined from Equation 3.32. The transition of the three-level system on the other hand must be determined by

$$\tau_{3\,level} = \tau_{phonon} + \tau_{triplet}. \tag{3.35}$$

And Equation 3.3 must be rewritten as

$$\frac{dn_2}{dt} = -A_{21}n_2 - \frac{n_2}{\tau_{3\,level}}. \tag{3.36}$$

Where the radiative lifetime for the system becomes

$$\tau = \frac{1}{A_{21} + \dfrac{1}{\tau_{3\,level}}}. \tag{3.37}$$

**Example 3.1: Atoms Trapped in the Triplet State**

Now consider the $N$ atoms in our box being uniformly excited by a beam of light containing enough photons to excite them all simultaneously. Also, the beam of light will stay on such that as an atom is de-excited it will be excited again. Assume our system of atoms in a box is designed in such a way that the probability of a singlet-state emission occurring is 90% and a triplet state 10%. Let's also assume our system is made of atoms that have the following radiative lifetimes:

$$\tau_{singlet} = 10\,ns, \tag{3.38}$$

$$\tau_{\text{phonon}} = 1\,\mu\text{s}, \qquad (3.39)$$

$$\tau_{\text{triplet}} = 9\,\mu\text{s}. \qquad (3.40)$$

In the first 10 ns, 90% of the atoms will emit a singlet-state transition photon and 10% of them will be started into the vibrational transition from singlet to triplet state. This will continue like so:

0–10 ns 90% singlet, 10% singlet to triplet vibrational state
10–20 ns 81% singlet, 19% singlet to triplet vibrational state
20–30 ns 73% singlet, 27% singlet to triplet vibrational state
30–40 ns 66% singlet, 34% singlet to triplet vibrational state
40–50 ns 59% singlet, 41% singlet to triplet vibrational state
50–60 ns 53% singlet, 47% singlet to triplet vibrational state
60–70 ns 48% singlet, 52% singlet to triplet vibrational state
70–80 ns 43% singlet, 57% singlet to triplet vibrational state
80–90 ns 39% singlet, 61% singlet to triplet vibrational state
and so on until about
~450 ns 1% singlet, 99% singlet- to triplet-vibrational state.

Now practically all of the atoms are trapped in the transition to triplet state and have yet to emit the triplet to ground-state photon! Also, the incident excitation beam is having no impact on the system since all of the atoms are already in an excited state and have yet to transition back to the ground state. Typically, this is how phosphorescence works. We can now turn off the incident light and the system will start to glow as the triplet-state electrons transition back to the ground state.

Example 3.1 is also important when it comes to the discussion of various types of lasers such as liquid organic dye lasers (discussed in Chapter 5). These types of lasers function exactly in this way and require intense short optical pulses to excite the dye molecules. The molecules have triplet states and after a period of microseconds to milliseconds (depending on the dye and the type of incident light) will shut off due to this triplet-state trapping. In fact, some of these types of lasers will have a chemical, element, or compound added to the dye solution that will help prevent the triplet state from occurring. These additives are called triplet-state *quenchers*.

## 3.3 A More Realistic Model

### 3.3.1 The Two- and Three-Level Models

Figure 3.10 shows the two-level and the three-level diagrams or models for absorption, spontaneous emission, stimulated emission, and *phonon* emission or vibrational states. The two-level system is the most simplistic of the models and is quite useful in understanding the basic idea of the *principle quantum number* and electron shell energy levels. We even used the two-level model in Figures 3.6–3.8 to demonstrate the spin quantum number and how it governs the singlet, doublet, or triplet state of the transition. It is also useful to think of the two-level system (actually an *N*-level system of singlet states) when thinking of how the ladder operators are used to move the energy level up or down a quanta as they are applied.

The three-level system shown in Figure 3.10 gives us a slightly more detailed model with more possible transition states. Here, we first introduced the idea of the *phonon* and the vibrational states representative of heat or mechanical transfer of energy that takes place in an ensemble of atoms or molecules. While the three-level model is closer to reality there are still a few pieces missing to it that we will need to add before our model is ready to explain all of the absorption and emission processes pertinent to our laser discussion.

### 3.3.2 The Four-Level Model

Figure 3.11 shows the four-level model that is a bit more realistic in that due to heat, mechanical vibrations, collisions, external electromagnetic fields, and other phenomena it is unlikely that a singlet-absorption transition occurs in which the electron jumps to the exact energy level necessary for it to transition back down to ground with a quanta of energy. The classical analog might be to assume all systems have losses like friction

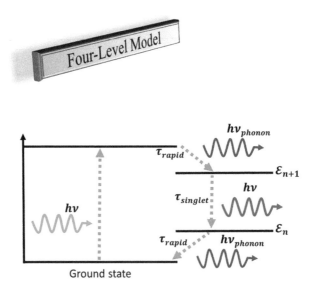

**FIGURE 3.11**
The four-level model.

and inefficiencies in energy transfer. In actuality, the quantum mechanical system has a myriad of other quantum numbers that are beyond the scope of this text to go into. Suffice it to say that there are many quantum leaps that can occur in an ensemble system such as our $N$ atoms in a box example.

In Figure 3.11, the four-level model shows that a photon is absorbed at some original input energy level (wavelength and/or frequency). There is some short vibrational transition to another singlet state where spontaneous or stimulated emission occurs to an energy level slightly above the ground state. The electron then emits another phonon for a vibrational transition all the way back to the ground state. We must note here that these vibrational transitions are not like the triplet-state transitions that might be long in duration. Instead, these are very rapid transitions and there is no fear of getting "trapped" as shown in Example 3.1. Figure 3.12 shows the four-level model with a pathway for triplet-state transitions shown. Again, the triplet state would be an undesirable situation for a laser system.

Figure 3.13 shows another interesting situation that is even more like the real world. Many ensemble systems with multiple molecules or atoms will have many vibrational modes and many singlet-excited states that can occur. This means that the system might emit photons at various energy levels and therefore it will not be a singular wavelength or color. The linewidth of the spectrum will be broader as it is a summation of many possible emissions. There are so many physical situations and interactions that can impact these transitions that there is no point to go into all of them. It is important for the laser scientist or engineer to understand that this occurs and that the spectral linewidth will be specific to each individual ensemble system scenario.

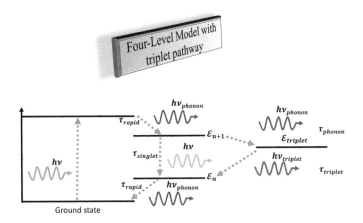

**FIGURE 3.12**
The four-level model with the possible triplet-state transition pathway shown.

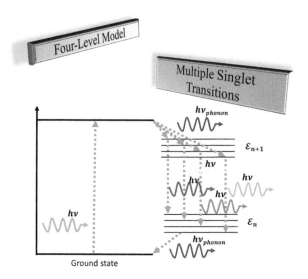

**FIGURE 3.13**
The four-level model with many possible singlet transitions.

## 3.4 Chapter Summary

In our quest to develop an understanding of what a laser actually is it makes sense that our third discussion following our discussions on light and then amplification should be the next major words in the acronym, laser itself, which is "stimulated emission of radiation." In this chapter, we discussed the details of how light interacts with atoms and electrons.

In Section 3.1, we discussed the Bohr model of the atom and how electrons can only make quantized leaps in energy levels from one atomic orbital or shell to the next. We then discussed *Pauli's Exclusion Principle* and singlet, doublet, and triplet states of the atom.

In Section 3.2, we developed rate equations to describe the interaction of light with atoms through absorption, spontaneous emission, and stimulated emission. We learned that the quantum mechanical harmonic oscillator and the ground state of the universe is a nonzero electromagnetic field and therefore is the trigger for spontaneous emission. We also developed a model of atoms with vibrational transition paths to triplet states and discussed how this was like some lasers today.

In Section 3.3, we looked at the models for absorption, spontaneous and stimulated emission, and vibrational states. We discussed the two-, three-, and four-level systems in some detail and even discussed how more detail can be added that will show us why some ensembles of atoms have multiple spectral lines and broad linewidths.

We now have a detailed understanding of what the modern theory of light is. We understand a bit more about amplifiers and amplification than we did at the start. And we now also understand what the "stimulated emission of radiation" is. We have developed all the tools needed to lead us into Chapter 4 where we will finally answer the question: "What are lasers?"

## 3.5 Questions and Problems

1. What is a "quantum leap"?
2. Draw a picture of the Bohr model of the atom.
3. What is the *principle quantum number?*
4. State the *Pauli Exclusion Principle.*
5. Realizing that

$$c = \lambda v \tag{3.41}$$

and the wavelength, $\lambda$, of the singlet-state spontaneous emission of an atom is 632 nm, what is the absorption energy required.
6. Define *singlet, doublet,* and *triplet states.*
7. What is the difference between fluorescence and phosphorescence?
8. Solve Equation 3.3 and show that time constant for the decaying model is the inverse of the Einstein A coefficient.
9. Explain what Equation 3.12 means physically.
10. Given that:

$$a^{\dagger}\psi_0 \equiv 0. \tag{3.42}$$

and

$$a^{\dagger}\psi_0 = 0 = \frac{1}{\sqrt{2m}}\left(\frac{\hbar}{j}\frac{\partial \psi_0}{\partial x} - jm\omega x\psi_0\right). \tag{3.43}$$

Show that the ground-state energy of the quantum harmonic oscillator is nonzero.
11. While Equation 3.28 is "unnecessary" and perhaps even nonsensical, explain why it was discussed and what philosophical importance it might hold for calling spontaneous emission stimulated emission instead.

12. Use Equation 3.30 to estimate the maximum amount of time a 500-nm virtual photon can exist in real space.

13. What is *radiative lifetime*?

14. In Equation 3.33, how many multiples of the *radiative lifetime* constant occur before the value of $n(t)$ is effectively zero? Hint: graph the equation.

15. What is a *phonon*?

16. Use Equation 3.37 in 3.33 and graph it for multiple values of $A_{21}$. Assume $\tau$ is 10 ns.

17. Write a computer program, application, or model in math software to simulate the $N$ atoms in a box as discussed in Example 3.1.

18. What is a four-level system? Draw it.

19. How can an ensemble of atoms in a box (all the same type of atom) generate emission with more than one singlet-state energy (wavelength or color)?

20. Draw an $N$-level singlet-state model to depict the ladder operators' functions.

# 4

## *What Are Lasers?*

As we discussed in Chapter 1, the word laser is an acronym that actually stands for Light Amplification by the Stimulated Emission of Radiation. We started by looking at the acronym itself as the simplest method to determine where to start in our study of the laser. The very first word in the acronym is "light." So, we discussed in great detail in Chapter 1 the answer to the question: "What is light?"

It makes sense that our next discussion, as given in Chapter 2, was the next major word in the acronym itself, which is "amplification." We discussed amplification and amplifiers in some detail. Now we can complete our analysis of the acronym. In Chapter 3 we asked the question, what is "stimulated emission of radiation?" We found that the way that light interacts with matter is through absorption and emission. An electron will either absorb light and move to an excited state or it will emit light and drop from the excited state to a lower-energy state. When an excited electron is exposed to an external field the "stimulated emission" occurs.

In this chapter, we will discuss the details of what laser actually means. What is a laser and how does it work? We now have the basic background information needed to put the concept together and begin our quest in becoming laser scientists and engineers.

## 4.1 Laser Basics

### 4.1.1 Components

Figure 4.1 shows the basic components of a laser. The key to the laser is the *active gain medium,* which is the component that contains the atoms, molecules, or other mechanisms (electrons, ions, solid-state "holes," etc.) that are used for stimulated emission. As an example, a Neodymium-doped yttrium orthovanadate ($Nd^{+3}$:$YVO_4$) laser uses the transparent $YVO_4$ crystal as a mechanical host medium, where neodymium ions are doped into the said host material. It is the neodymium ions that can absorb and emit photons through the stimulated emission process. The *active gain medium* can be considered to be a matrix of elements that will provide for stimulated emission. The density of these elements in the medium plays a key role in the output power of the laser system.

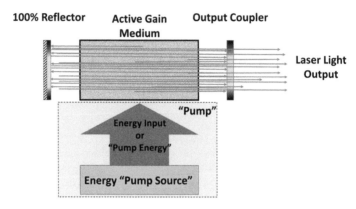

**FIGURE 4.1**
The basic components of the laser.

Recall Figure 2.25 with the many series amplifiers and feedback loop. The system given in Figure 2.25 is very similar to that shown in Figure 4.1. The active gain medium can be represented by the amplifiers. The feedback loop is physically implemented in a laser by a 100% reflective mirror, or *reflector*, on one end of the optical cavity and a partially reflective *output coupler* on the other end. Based on the percentage of reflectance of the *output coupler* mirror a certain number of photons are reflected back into the cavity with each pass through and some are allowed to escape.

Of course, there would be no stimulated emission if there is no external energy input into the system. This energy input comes from an external *energy source* and uses some mechanism to convert energy (and therefore power) from a source into a form of energy that the *active gain medium* can absorb and become excited. This energy input is also called the "pump energy" that comes from the "pump source." Often the two components are discussed in combination simply as the "pump."

For example, in the case of the $Nd^{+3}:YVO_4$ laser the "pump" comes from a diode laser that emits near infrared light at a wavelength of about 808 nm. The laser light is absorbed by the Nd ions exciting them to a higher-energy level. There is a phonon decay to a singlet state where an infrared 1,064-nm photon is released. The ion then releases another phonon and relaxes to the ground energy level. The laser process for $Nd^{+3}:YVO_4$ is shown in the four-level energy diagram in Figure 4.2. There are other transitions at other wavelengths that can occur but the laser optics are typically designed in such a way as to only allow the 1,064 nm to reflect back and forth and pass through the cavity and therefore generate stimulated emission. We say this is a "selected" transition due to the physical design of the laser system.

We should also note here that in general with lasers the input light could come from flashlamps, electric arcs, and other lasers. The idea is to pump the active gain medium at or near the peak of its material absorption band so it

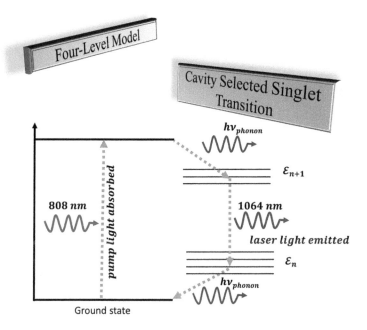

**FIGURE 4.2**
The four-level energy diagram for the YV:O$_4$ laser.

will absorb the input energy efficiently. The *active gain medium* then becomes excited.

Figure 4.3 shows the ensemble of amplifiers from Figure 2.25 compared to the laser components in Figure 4.1. The similarities exist because the amplifier model actually is the model for the laser in the frequency domain. It's a good thing we went over that in Chapter 2! Let's revisit that discussion from a different point of view.

To summarize this section it is good to put into list form the major components of a laser. Those components are:

1. *active gain medium*
2. pump source (source of power)
3. pump mechanism (method for inputting power into the medium)
4. optical cavity which includes
   a. 100% reflector
   b. Partial reflective/transmissive output coupler
   c. Other optics per specific laser design needs.

There are other subcomponents of lasers for mechanical-, electrical-, and thermal-engineered purposes such as coolers with flowing coolants,

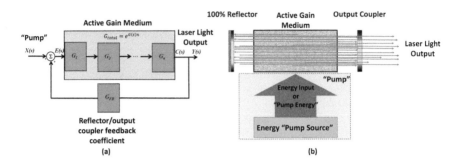

**FIGURE 4.3**
(a) The frequency-domain amplifier model of the laser compared to (b) the components of the laser.

mechanically stable structures, high-voltage circuits, and more. Lasers have many components and can become quite complicated in actuality even though Figure 4.1 seems rather simple.

### 4.1.2 Gain, Population Inversion, and Saturation

#### 4.1.2.1 Gain

In Chapter 2, we considered an ensemble of amplifiers with a feedback loop as shown in Figure 4.3a. Now we will follow the discussion that started with Equation 2.29 but with Figure 4.3b in mind.

Consider the gain medium as a matrix of particles that can go through the process of stimulated emission as discussed in Chapter 3. Inside the active medium, light is initiated through a vacuum energy fluctuation-triggered spontaneous emission, which we called an input signal in Chapter 2. This initial photon then propagates linearly along the z-axis and is amplified each time it triggers stimulated emission. The one photon becomes two, two becomes four, four becomes eight, and so on creating an "avalanche" of photons. The more and more the light signal passes through the cavity, the more the signal is amplified through stimulated emission as shown in Figure 4.4. Therefore, an equation describing the overall amplification of the irradiance as it passes through the cavity can be written in terms of the gain coefficient, $G(\omega)$, and propagation distance, $z$, as

$$\frac{dI(z)}{dz} = G(\omega)I(z). \tag{4.1}$$

Rearranging the equation results in

$$\frac{dI(z)}{I(z)} = G(\omega)dz. \tag{4.2}$$

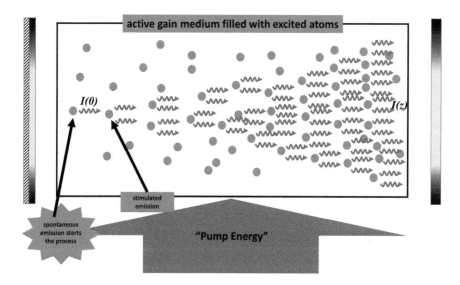

**FIGURE 4.4**
The active gain medium enables an "avalanche" of photon generation through stimulated emission.

Integrating Equation 4.2 gives

$$\ln\left|\frac{I(z)}{I(0)}\right| = G(\omega)z \qquad (4.3)$$

or

$$I(z) = I(0)e^{G(\omega)z}. \qquad (4.4)$$

Figure 4.4 only depicts one pass through the laser's *active gain medium,* but based on the reflectance of the mirrors at the ends many of the photons are reflected back into the cavity and therefore continue the gain process. Now let's consider multiple passes through the laser shown in Figure 4.4. We will assume the active gain medium to be $L$ long and since there is a 100% reflector on one end the maximum path for gain to occur is $2L$. Therefore, Equation 4.4 can be written for the first pass of $n = 1$ as

$$I(n = 1) = I_0 e^{G2L}. \qquad (4.5)$$

For simplicity we will write $I(0) = I_0$ and $I(z = 2L) = I(n)$. We now must realize that the output coupler mirror lets some percentage of the light out of the cavity but also reflects some. Letting the reflectance of the output coupler be $R_c$ we then can write the equation for pass $n = 2$ as

$$I(n = 2) = R_c I_0 e^{G2L} e^{G2L} = R_c I_0 e^{G4L}. \qquad (4.6)$$

The result of Equation 4.6 is now $I_0$ for the next pass $n = 3$, which is

$$I(n = 3) = R_c R_c I_0 e^{G4L} e^{G2L} = R_c^2 I_0 e^{G6L}. \tag{4.7}$$

Likewise, pass $n = 4$ is

$$I(n = 4) = R_c R_c^2 I_0 e^{G6L} e^{G2L} = R_c^3 I_0 e^{G8L}. \tag{4.8}$$

Inspection of this process from pass 1 to 4 allows us to write a general equation for the maximum gain based on the number of passes through the cavity, $n$, as

$$I(n) = R_c^{n-1} I_0 e^{2nGL}. \tag{4.9}$$

Figure 4.5 shows Equation 4.9 graphically while using real numbers from a typical $Nd^{+3}:YVO_4$ laser pointer-type active medium where $L = 0.01\,cm$, $G = 315\,cm^{-1}$, $R_c = 0.5$, and $I_0 = 1\,\mu W/cm^2$. Examination of Equation 4.9 and Figure 4.5 suggests that as long as we continue to let the laser photons pass back and forth between the cavity mirrors we would eventually achieve an infinite irradiance output from the laser! In fact, Figure 4.5 shows more output power than from a nuclear blast in just a dozen passes through the cavity!

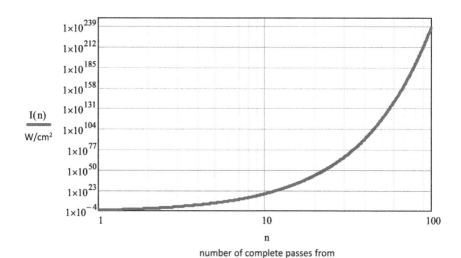

**FIGURE 4.5**
Irradiance gain as a function of the number of passes through an $Nd^{+3}:YVO_4$ laser-active medium with $L = 0.01\,cm$, $G = 315\,cm^{-1}$, $R_c = 0.5$, and $I_0 = 1\,\mu W/cm^2$.

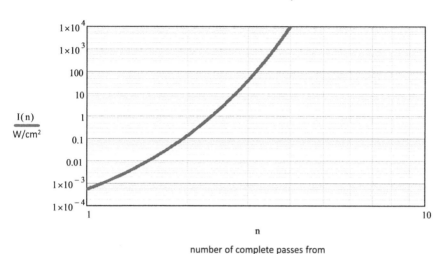

n

number of complete passes from
output coupler to reflector and back

**FIGURE 4.6**
Irradiance gain as a function of the number of passes through an $Nd^{+3}$:$YVO_4$ laser-active medium with $L = 0.01$ cm, $G = 315$ cm$^{-1}$, $R_c = 0.5$, and $I_0 = 1$ μW/cm$^2$.

One would think that something is clearly wrong with this model. In actuality, we must realize that the model is really only good for a few passes because there are other phenomena that occur within the cavity that become detrimental to the system's gain. Figure 4.6 shows the graph from Figure 4.5, but zoomed in to only three passes within the cavity and we see that the amount of irradiance is already approaching levels that would actually damage the crystals and optics within the laser. In fact, many lasers only ever need or achieve more than a single to a few passes within the cavity.

### 4.1.2.2 Population Inversion

Reconsider Equation 4.4 rewritten simply as

$$I(z) = I_0 e^{Gz}. \tag{4.10}$$

At this point, we need to reexamine our *active gain medium* as shown in Figure 4.7. Assume that all the atoms of the material are in the ground state. In other words, the density of atoms in the material in the ground state is $N_1$ atoms/cm$^3$. If a beam of light is passed into the medium from one side to the other, or $z = 0$ to $L$ then that material will absorb some of this light and some of these atoms will be excited to the next energy level as discussed in Chapter 3. The material's likelihood of absorbing the incident light is described by the *cross section coefficient for absorption*, $\sigma_{12}$ which is measured in units of area (cm$^2$ typically).

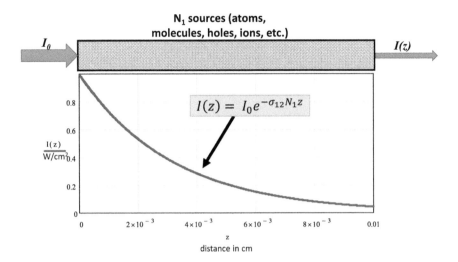

**FIGURE 4.7**
Irradiance absorption as a function of the distance, $z$, through an $Nd^{+3}{:}YVO_4$ laser-active medium with $L = 0.01$ cm, $\sigma_{12} = 25 \times 10^{-19}$ cm$^{-1}$, $N_1 = 1.26 \times 10^{20}$ atoms/cm$^3$, and $I_0 = 1$ W/cm$^2$.

The gain, $G$, for the active medium is

$$G = \sigma_{12} N_1. \qquad (4.11)$$

In the case of absorption, the gain is negative so we can rewrite Equation 4.10 for absorption of an input irradiance through the medium as

$$I(z) = I_0 e^{-\sigma_{12} N_1 z}. \qquad (4.12)$$

Figure 4.7 shows Equation 4.12 graphed with parameters based on the $Nd^{+3}{:}YVO_4$ laser pointer from our previous discussion above. As the light passes through the medium it is absorbed by the neodymium ions and therefore exciting them. The *cross section coefficient for absorption*, $\sigma_{12}$ is the same as the *cross section coefficient for emission*, $\sigma_{21}$. So, if all the atoms in the medium are in the excited state initially then we have a population density of $N_2$ atoms/cm$^3$ in the material and Equation 4.12 can be written as

$$I(z) = I_0 e^{\sigma_{21} N_2 z}. \qquad (4.13)$$

Figure 4.8 shows the gain through the medium based on Equation 4.13 and the $Nd^{+3}{:}YVO_4$ laser pointer parameters.

Realizing that $\sigma = \sigma_{12} = \sigma_{21}$ and, if there is some population density of atoms, $N_1$, in the ground state, and some population density, $N_2$, in the excited state then Equation 4.13 becomes

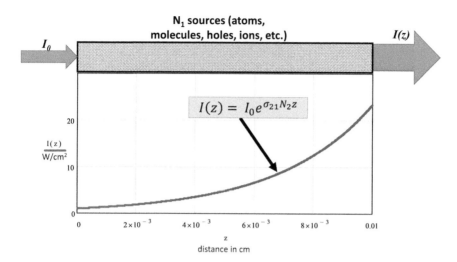

**FIGURE 4.8**
Irradiance gain as a function of the distance, $z$, through an $Nd^{+3}{:}YVO_4$ laser-active medium with $L = 0.01$ cm, $\sigma_{12} = 25 \times 10^{-19}$ cm$^{-1}$, $N_1 = 1.26 \times 10^{20}$ atoms/cm$^3$, and $I_0 = 1$ W/cm$^2$.

$$I(z) = I_0 e^{\sigma(N_2 - N_1)z}. \tag{4.14}$$

For situations when

$$\frac{N_2}{N_1} < 1 \tag{4.15}$$

the medium will be in the absorbing mode. For situations when

$$\frac{N_2}{N_1} > 1 \tag{4.16}$$

there will be more atoms in the excited state than in the ground state which is called a *population inversion* and the gain will be positive as the input beam passes through the cavity. A significant population inversion is required for a laser to operate.

### 4.1.2.3 Saturation

We must recall that lasers are amplifiers. In fact, the very operation of the laser relies on amplification as has already been discussed in detail in Chapters 2 and 3. As discussed in Section 2.1.2 at some point an amplifier can no longer increase the signal amplitude as it reaches the power available limit. This limit is known as *saturation*.

Consider the $Nd^{+3}{:}YVO_4$ laser pointer previously discussed. This laser has a small block of crystal about 3 mm by 3 mm by 3 mm in cubic size.

Throughout the YVO$_4$ crystal there are about $2 \times 10^{18}$ Nd$^{+3}$ ions doped homogenously within it. The output irradiance, $I_{laser}$, will be at a maximum output when every single neodymium ion is excited and going through the stimulated emission process. In other words, the maximum output of the laser at a wavelength, $\lambda$, of 1,064 nm is

$$I_{laser} = n \frac{hc}{\lambda} \frac{1}{A} \qquad (4.17)$$

where $n$ is the number of photons in the cross-sectional area, $A$, of the laser beam profile. For a laser pointer the beam is typically a round spot about 1 mm in radius. Therefore,

$$I_{laser} = n \frac{hc}{\lambda \pi r^2}. \qquad (4.18)$$

Nd$^{+3}$:YVO$_4$ laser pointers are diode pumped by an input laser source of 808 nm and each of the pump laser photons have about a 60% chance of being converted to a stimulated emission photon at the 1,064-nm wavelength. This percentage is known as the *conversion efficiency*, $\eta$, of the laser and is found as

$$\eta = \frac{I_{pump}}{I_{laser}} = \frac{n_{pump} \dfrac{hc}{\lambda_{pump} \pi r^2}}{n_{laser} \dfrac{hc}{\lambda_{laser} \pi r^2}} = \frac{n_{pump} \lambda_{laser}}{n_{laser} \lambda_{pump}}. \qquad (4.19)$$

It is similar to another quantity often used to describe laser efficiency which is the *quantum defect* of the laser, $q$,

$$q = I_{pump} - I_{laser} = n_{pump} \frac{hc}{\lambda_{pump} \pi r^2} - n_{laser} \frac{hc}{\lambda_{laser} \pi r^2}. \qquad (4.20)$$

Figure 4.9 shows a graph of the pump irradiance and the laser output irradiance as a function of the number of photons as given in Equations 4.17 and 4.18. Once the laser medium has the total number of neodymium ions in the excited state the output plateaus and the laser becomes *saturated*. Once the amplifying entities (neodymium ions in this case) are all excited no matter how much input pump power there is the laser output will no longer increase. This is the case of full gain medium saturation throughout the volume of the laser cavity.

We should also point out that there are other interesting saturation phenomena when it comes to lasers. A single spatial or volumetric region of the gain medium might become saturated while other parts of the medium do not. This is called *spatial hole burning* and has interesting ramifications to the laser's spatial output beam profile. It is typically caused by two laser modes created within the medium that propagate in opposite directions of

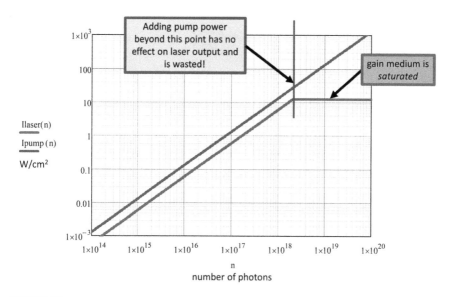

**FIGURE 4.9**
Once all the gain medium-stimulated emission entities (such as atoms, ions, holes, etc.) become excited the medium becomes saturated.

each other. The interaction of these two counter-propagating beams saturates the medium along their overlapping paths.

It is also possible for a mode of laser operation within the medium to saturate the gain at a specific wavelength. Some gain media have a broad emission spectrum but due to the optics arrangement of the laser cavity a specific transition is selected as discussed in Section 4.1.1. The specific "selected" transition has a much narrower linewidth than the gain medium's spectrum and if the particular transition is pumped hard enough or has a narrow linewidth input beam with large enough irradiance the gain at that wavelength can become saturated while gain at others cannot. This is called *spectral hole burning* of the medium.

## 4.2 The Rate Equation

### 4.2.1 The Four-Level Laser

We discussed the four-level transition model in Chapter 3 in some detail. Now we need to apply that same model to more than just one atom or excitation entity. Figure 4.10 shows the four-level transition model for a population of $N$ atoms. The individual atoms each has four energy levels in which it can exist in represented by the subscripts 1–4. Therefore, the

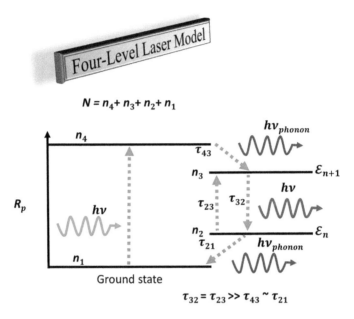

$$N = n_4 + n_3 + n_2 + n_1$$

**FIGURE 4.10**
The four-level laser model is used to develop the laser rate equations.

population of atoms will consist of some number of atoms in each of the states. In other words,

$$N = n_4 + n_3 + n_2 + n_1. \tag{4.21}$$

As shown in Figure 4.10, $n_1$ is the ground state, $n_4$ is the upper level, $n_3$ is the upper laser level, and $n_2$ is the lower laser level. The pump source excites the atoms to the upper level with a rate, $R_p$. The atom quickly relaxes to the upper laser level with a transition time, $\tau_{43}$. The major portion of the population will be trapped in the upper laser level because the transition time, $\tau_{32}$, is much longer than $\tau_{43}$ and $\tau_{21}$. This means that the total number of atoms in the population can be approximated as being $n_3$ until it decays to $n_2$. So, we will omit calculations for atoms in the $n_4$ and $n_1$ state at this point as they are not where all the laser action is!

The change in the population of $n_3$ is written as

$$\frac{dn_3}{dt} = R_p - B_{\text{laser}}\rho(v)(n_3 - n_2) - \frac{n_3}{\tau_{32}}. \tag{4.22}$$

In the above equation $B_{\text{laser}} = B_{32} = B_{23}$ is the Einstein coefficient for stimulated emission and absorption transitions, respectively, and $\rho(v)$ is the spectral energy density field as discussed in Chapter 3. The change in population in $n_2$ is

$$\frac{dn_2}{dt} = B_{\text{laser}} \rho(v)(n_3 - n_2) + \frac{n_3}{\tau_{32}} - \frac{n_2}{\tau_{21}}. \qquad (4.23)$$

Letting $W_{\text{laser}} = B_{\text{laser}}\rho(v)$ be the transition probability for stimulated emission allows us to rewrite the two rate equations as

$$\frac{dn_3}{dt} = R_p - W_{\text{laser}}(n_3 - n_2) - \frac{n_3}{\tau_{32}} \qquad (4.24)$$

and

$$\frac{dn_2}{dt} = W_{\text{laser}}(n_3 - n_2) + \frac{n_3}{\tau_{32}} - \frac{n_2}{\tau_{21}}. \qquad (4.25)$$

At this point, we will introduce a decay rate coefficient, $\gamma = 1/\tau$ to simplify the next few steps. Rewriting Equations 4.24 and 4.25 with this new decay rate

$$\frac{dn_3}{dt} = R_p - W_{\text{laser}}(n_3 - n_2) - \gamma_{32} n_3 \qquad (4.26)$$

and

$$\frac{dn_2}{dt} = W_{\text{laser}}(n_3 - n_2) + \gamma_{32} n_3 - \gamma_{21} n_2. \qquad (4.27)$$

The steady-state solution to the rate equations can be found by realizing that the rate of change in each must be equal to zero. Solving the two simultaneous equations results in

$$n_3 = R_p \frac{W_{\text{laser}} + \gamma_{21}}{W_{\text{laser}} \gamma_{21} + \gamma_{21} \gamma_{32}} \qquad (4.28)$$

and

$$n_2 = R_p \frac{W_{\text{laser}} + \gamma_{32}}{W_{\text{laser}} \gamma_{21} + \gamma_{21} \gamma_{32}}. \qquad (4.29)$$

At this point, we need to introduce another rate equation that will tell us the number of photons produced inside the laser cavity. We will define $\phi(t)$ as the total number of photons in the cavity and it is found to be

$$\frac{d\phi}{dt} = V_a B N_3 (\phi + 1) - \frac{\phi}{\tau_c}. \qquad (4.30)$$

In Equation 4.30, $V_a$ is the volume the optical field fills within the active medium, $B$ is the stimulated transmission rate per photon per mode, where the "mode" is the particular optical field of particular wavelength. $N_3$ is the total number of atoms in the laser transition excited state. And $\tau_c$ is the photon lifetime. Note that the (...+1) portion is to represent an initial photon in the cavity created via spontaneous emission. Since $\phi$ will be some value much larger than 1 we can simplify Equation 4.30 as

$$\frac{d\phi}{dt} = \left( V_a B N_3 - \frac{1}{\tau_c} \right)\phi. \tag{4.31}$$

Consider Equation 4.14 for the four-level laser with active medium of length, $l$, shown in Figure 4.11 where

$$I = I_0 e^{\sigma N_3 2l}. \tag{4.32}$$

Note the $2l$ is due to the beam starting at $R_2$ mirror then hitting $R_1$ and back again having traveled $2l$ through the active medium (we consider this one pass through the cavity). We can write the value for the irradiance in the cavity, $I_1$ after the first pass as a function of losses in the cavity

$$I_1 = I_0 R_1 R_2 \left( 1 - L_i \right)^2 e^{\sigma N_3 2l}. \tag{4.33}$$

$R_1$ and $R_2$ are the reflectances of the two cavity mirrors and $L_i$ is the loss per single pass within the laser medium. We can rewrite the reflectances in terms of internal mirror loss or inefficiency, $a_1$ and $a_2$, and the transmittances of the mirrors, $T_1$ and $T_2$ as

$$R_1 = 1 - a_1 - T_1 \tag{4.34}$$

and

$$R_2 = 1 - a_2 - T_2. \tag{4.35}$$

Realizing that $a_1$ and $a_2$ are small and likely the same material for each mirror (we will assume they are) and therefore can both be approximated

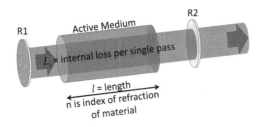

**FIGURE 4.11**
Laser cavity with active medium of length, l, and index of refraction, $n$, mirrors with reflectivities, $R_1$ and $R_2$, and internal loss, $L_i$.

as simply $a$ and using the approximation $(1-a-T) \sim (1-a)(1-T)$ then we can rewrite Equation 4.33 as

$$\Delta I = \left[ (1-T_1)(1-T_2)(1-a)^2 (1-L_i)^2 e^{\sigma N_3 2l} - 1 \right] I. \tag{4.36}$$

As it turns out the loss internal to the laser and at the mirrors is typically found to be logarithmic and can be written as

$$\gamma_1 = -\ln(1-T_1) \tag{4.37}$$

$$\gamma_2 = -\ln(1-T_2) \tag{4.38}$$

$$\gamma_{cav} = -\left[ \ln(1-a) + \ln(1-L_i) \right]. \tag{4.39}$$

Using Equations 4.37–4.39 we can rewrite Equation 4.36 as

$$\Delta I = \left[ e^{-\ln(1-T_1)} e^{-\ln(1-T_2)} e^{-2[\ln(1-a)+\ln(1-L_i)]} e^{\sigma N_3 2l} - 1 \right] I \tag{4.40}$$

or

$$\Delta I = \left[ e^{\gamma_1} e^{\gamma_2} e^{-2\gamma_{cav}} e^{\sigma N_3 2l} - 1 \right] I. \tag{4.41}$$

Simplifying Equation 4.41

$$\Delta I = \left[ e^{-[\gamma_1+\gamma_2+2\gamma_{cav}]} e^{\sigma N_3 2l} - 1 \right] I \tag{4.42}$$

$$\Delta I = \left[ e^{-2\left[\gamma_{cav}+\frac{\gamma_1+\gamma_2}{2}\right]} e^{\sigma N_3 2l} - 1 \right] I \tag{4.43}$$

We will define a total cavity loss as

$$\gamma = \gamma_{cav} + \frac{\gamma_1+\gamma_2}{2}. \tag{4.44}$$

Therefore, Equation 4.43 becomes

$$\Delta I = \left[ e^{2(\sigma N_3 l - \gamma)} - 1 \right] I. \tag{4.45}$$

Using a Taylor series expansion for the exponential in Equation 4.45

$$\Delta I \cong (1+2(\sigma N_3 l - \gamma)-1) I \tag{4.46}$$

$$\Delta I \cong (2\sigma N_3 l - \gamma) I. \tag{4.47}$$

Divide both sides of Equation 4.47 by the time taken for a single pass through the cavity $\Delta t$. Therefore,

$$\frac{\Delta I}{\Delta t} \cong \frac{1}{\Delta t}(2\sigma N_3 l - \gamma)I. \tag{4.48}$$

Realizing that

$$\Delta t = \frac{2L'}{c} \tag{4.49}$$

where $L'$ is the optical path length of the cavity, $c$ is the speed of light in the vacuum, and that the left-hand side is approximately the differential of the irradiance with respect to time, then Equation 4.48 becomes

$$\frac{dI}{dt} = \left(\frac{\sigma N_3 lc}{L'} - \frac{\gamma c}{L'}\right)I. \tag{4.50}$$

Recall Equation 4.31 and compare it to Equation 4.50. We defined $\phi(t)$ as the total number of photons in the cavity and we have also learned that the irradiance is the number of photons multiplied by the quantized energy, $h\nu$, divided by the beam area. So, it becomes clear that the portion of Equation 4.50 in the parentheses is the same as the portion of Equation 4.31 in parentheses, or,

$$V_a BN_3 - \frac{1}{\tau_c} = \frac{\sigma N_3 lc}{L'} - \frac{\gamma c}{L'}. \tag{4.51}$$

From inspection we see that

$$V_a BN_3 = \frac{\sigma N_3 lc}{L'} \tag{4.52}$$

and

$$\frac{1}{\tau_c} = \frac{\gamma c}{L'}. \tag{4.53}$$

We will now rewrite Equation 4.44 as

$$\gamma = \frac{L'}{\tau_c c} = \gamma_{cav} + \frac{\gamma_1}{2} + \frac{\gamma_2}{2}. \tag{4.54}$$

Simplifying Equation 4.54 in terms of the photon lifetime results in the following

$$\frac{1}{\tau_c} = \frac{c}{L'}\gamma_{cav} + \frac{c}{L'}\frac{\gamma_1}{2} + \frac{c}{L'}\frac{\gamma_2}{2}.$$ (4.55)

The far right argument of Equation 4.55 is the rate for the loss of photons out of the output coupler mirror 2. In other words, the power output of the laser is found by

$$P_{out} = \frac{c}{L'}\frac{\gamma_2}{2}h\nu\phi.$$ (4.56)

And the irradiance out of the laser is therefore

$$I_{out} = \frac{P_{out}}{A} = \frac{c}{AL'}\frac{\gamma_2}{2}h\nu\phi$$ (4.57)

where $A$ is the area of the output laser spot.

We can improve Equation 4.57 a bit by realizing some practical information about laser construction. Typically, there is some distance between the laser gain medium and the mirrors that might be empty space or open air. Since the gain medium has a different index of refraction, $n$, than the open air or empty space $L'$ is actually slightly longer than the actual distance between the laser mirrors, $L$, by a distance of $(n-1)l$ where $l$ is the length of the active medium. Equation 4.57 thus becomes

$$I_{out} = \frac{c\gamma_2}{2A(L+(n-1)l)}h\nu\phi.$$ (4.58)

We must point out here that the terms for the losses in Equations 4.37–4.39 are usually considered logarithmic for values if they are much larger than 0.25 or so, but for sufficiently high-reflectance mirrors, we can approximate them as

$$\gamma_1 = -\ln(1-T_1) \cong T_1$$ (4.59)

$$\gamma_2 = -\ln(1-T_2) \cong T_2$$ (4.60)

$$\gamma_{cav} = -[\ln(1-a)+\ln(1-L_i)] \cong a+L_i.$$ (4.61)

Figure 4.12 shows the loss term and the approximate term from Equation 4.59 as a function of $T_1$ for comparison. Figure 4.13 shows the percent difference between the two. Note that even at $T_1 = 0.50$ the percent difference is still only 16% so it is still somewhat reasonable of an approximation even at that transmittance. Certainly, the approximation is good for values of $T$ less than 0.3.

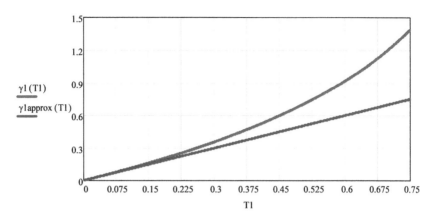

**FIGURE 4.12**
The logarithmic loss term compared to the approximate loss term.

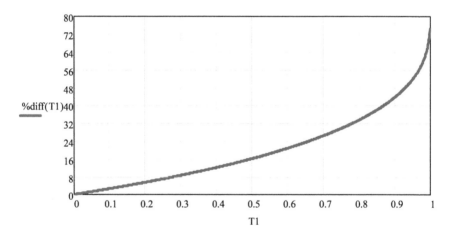

**FIGURE 4.13**
The percentage difference between the logarithmic and approximate loss terms.

Using the approximation in Equation 4.60, we can rewrite the irradiance output of the laser as a function of the output mirror transmission as

$$I_{out} = \frac{c}{2A\left(L+(n-1)l\right)} h\nu\phi T_2. \tag{4.62}$$

## 4.3 The Cavity

To this point, we have used the word "cavity" and sometimes the phrase "laser cavity" without truly discussing what they really are. The cavity is the optical system that causes the laser to perform with the multiple passes desired or to have a particular type of optical beam performance required. Other phrases that mean the same thing that the laser scientist or engineer might come across are optical cavity, resonator, resonating cavity, optical resonator, and oscillator or oscillator cavity. The cavity is an arrangement of optical components that might include reflective, refractive, dispersive, or diffractive properties.

Mirrors are the most common components and can be as close to 100% reflective as physically realizable down to single-digit percentages varying on specific cavity requirements. Most often, these mirrors are designed to reflect specific wavelengths or modes of light. Sometimes they are made with actual holes in them but typically the reflectance/transmittance percentages are accomplished by thin-film coatings.

Refractive elements such as lenses might be part of the cavity. These types of optical components are introduced into the laser cavity for various reasons. Lenses might be required to "collimate" a beam more concisely through the active medium or to reduce or increase the beam's divergence. With any focusing optics in laser operations, it is necessary to be mindful of the power within the laser cavity and the length of the laser pulse, because focusing down a high-power beam might cause ionization to occur in the active medium or any air gaps within the resonator configuration. Such material breakdowns can actually induce a plasma that absorbs all the laser light and shuts down laser operation. These types of breakdowns can also damage the active gain medium host.

Dispersive elements such as prisms are used to separate laser wavelength modes and to enable narrow linewidth operation. Just as a prism can separate sunlight into the colors of the rainbow, they are used in lasers to force specific desired wavelength transitions in the active gain medium.

Diffractive elements might be aperture stops (just a hole of specific size) to limit physical pathways between optics or they might be of specific design and purpose (such as a diffraction grating). Gratings are typically used much in the same way as prisms to select or narrow the linewidth of the laser oscillator.

### 4.3.1 Some Common Cavity Configurations

The most common cavity used in lasers is the *plane-parallel* resonator as shown in Figure 4.14. It consists of two flat reflective mirrors at each end. The radius of curvature, $R$, is infinitely long so that the mirrors are indeed mathematically flat.

Figure 4.15 shows a *confocal* cavity that consists of two spherical mirrors with equal radii of curvature of the same length as the optical path length between the mirrors. The beam between the two spherical mirrors has a minimum at the mid-distance point between them. This point is often referred to as the "waist of the beam."

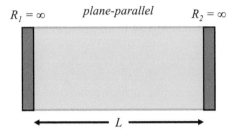

**FIGURE 4.14**
The plane-parallel laser cavity has two mirrors each with radius of curvature equal to infinity.

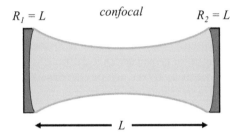

**FIGURE 4.15**
The confocal laser cavity has two mirrors each with radius of curvature equal to the optical path length of the cavity.

Similar to the *confocal* is the *spherical* cavity. The spherical resonator has two mirrors with equal radii of curvature set to one-half the length of the cavity as shown in Figure 4.16. Again, this cavity has a specific "waist" at the midpoint of the laser.

Figure 4.17 shows a hemispherical cavity where one mirror has a radius of curvature equal to the cavity length and the other is a flat mirror with radius equal to infinity. This one is similar to the *concave-convex* cavity shown in Figure 4.18. The *concave-convex* resonator has one *concave* mirror with radius

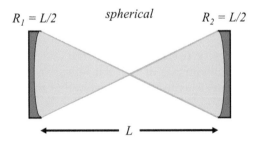

**FIGURE 4.16**
The spherical laser cavity has two mirrors each with radius of curvature equal to half the cavity length.

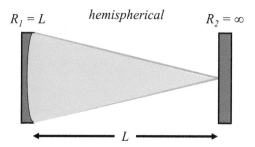

**FIGURE 4.17**
The hemispherical laser cavity has two mirrors one with radius of curvature equal to infinity and the other equal to the length of the cavity.

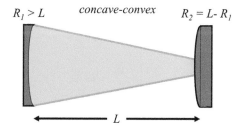

**FIGURE 4.18**
The concave–convex laser cavity has two mirrors: one concave mirror with radius of curvature greater than the length of the cavity and the other mirror is convex with radius of curvature equal to the length of the cavity minus the radius of the other mirror.

of curvature greater than the length of the cavity and the other mirror is *convex* with radius of curvature equal to the length of the cavity minus the radius of the other mirror.

There are many other cavity configurations with multiple components. In fact, there are so many different laser cavity configurations that entire books could have been written on the subject. We will not go into further detail on different configurations here; instead, we will strive to learn a bit more detail about the more common cavities as mentioned in this section.

## 4.3.2 Cavity Stability

Now that we have seen several optical cavity configurations, we need to discuss the idea of *stability*. An optical cavity is a *stable* cavity if the light traveling through it stays inside the geometric boundaries of it. The situations for *stability* of a cavity can be determined by tracing all the possible rays of light through it from end to end.

Consider the *confocal* cavity shown in Figure 4.15 where there are two flat mirrors on each end of the laser resonator separated by some distance, *L*. Each of the mirrors has a radius of curvature, *R*, but for now, we will assume

$R$ is unknown. (Note: be careful in this section when using $R$ as it can sometimes be radius of curvature and sometimes be reflectance of a mirror!) In
other words, our cavity might be *plane-parallel*, *confocal*, or *spherical*.

From classical ray tracing optics we can write a matrix for the cavity of
curved mirrors with focal lengths, $f = R/2$, as

$$\mathbf{M} = \begin{pmatrix} 1 & L \\ \dfrac{-1}{f} & 1 - \dfrac{L}{f} \end{pmatrix}. \tag{4.63}$$

In order to determine if the optical cavity is stable we must set up an eigenvalue equation using a factor $\lambda$ which can be real or complex. In order to
trace all of the "eigenrays" through the system we must realize that there is
a position equation in, $x$, and an angular position equation in $\theta$. So, the rays
through the system are described as

$$\mathbf{M} \begin{pmatrix} x_1 \\ \theta_1 \end{pmatrix} = \begin{pmatrix} x_2 \\ \theta_2 \end{pmatrix} = \lambda \begin{pmatrix} x_1 \\ \theta_1 \end{pmatrix}. \tag{4.64}$$

Using the identity matrix, $\mathbf{I}$, Equation 4.64 can be rewritten as the more
common eigenvalue equation

$$[\mathbf{M} - \lambda \mathbf{I}] \begin{pmatrix} x_1 \\ \theta_1 \end{pmatrix} = 0. \tag{4.65}$$

We take the determinate of the transfer matrix

$$\det[\mathbf{M} - \lambda \mathbf{I}] = 0 \tag{4.66}$$

$$\lambda^2 - \left(1 + 1 - \frac{L}{f}\right)\lambda + \det[\mathbf{M}] = 0 \tag{4.67}$$

$$\lambda^2 - \left(2 - \frac{L}{f}\right)\lambda + \left(1 - \frac{L}{f} - \frac{-L}{f}\right) = 0 \tag{4.68}$$

$$\lambda^2 - \left(2 - \frac{L}{f}\right)\lambda + \left(1 - \frac{L}{f} - \frac{-L}{f}\right) = 0 \tag{4.69}$$

$$\lambda^2 - \left(2 - \frac{L}{f}\right)\lambda + 1 = 0. \tag{4.70}$$

We will define the parameter, $g$, as the *stability parameter* equal to

$$g = 1 - \frac{L}{2f}. \tag{4.71}$$

Equation 4.70 can be rewritten as

$$\lambda^2 - 2g\lambda + 1 = 0. \tag{4.72}$$

The eigenvalues of Equation 4.72 are easily determined by

$$\lambda = g \pm \sqrt{g^2 - 1}. \tag{4.73}$$

Now consider a ray passing through the cavity the Nth time. The ray trace equation is

$$\begin{pmatrix} x_N \\ \theta_N \end{pmatrix} = \lambda^N \begin{pmatrix} x_1 \\ \theta_1 \end{pmatrix}. \tag{4.74}$$

$\lambda^N$ cannot grow infinitely large if the rays are to remain inside the cavity and therefore be stable. What we must realize here is that $\lambda^N$ must be periodic and bound by the cavity geometry (we will say that boundary is the absolute value of 1). We can rewrite Equation 4.74 as

$$\begin{pmatrix} x_N \\ \theta_N \end{pmatrix} = e^{\pm i\phi} \begin{pmatrix} x_1 \\ \theta_1 \end{pmatrix}. \tag{4.75}$$

where $\phi$ is an arbitrary phase factor. This also means that

$$0 \leq \lambda_+ \lambda_- \leq 1 \tag{4.76}$$

for stability. Substituting Equation 4.73 into 4.76 results in

$$0 \leq g_1 g_2 \leq 1 \tag{4.77}$$

where the subscripts correspond to which end of the cavity. Rewriting Equation 4.71 in terms of mirror curvature and cavity length and substituting it into Equation 4.77 yields the *cavity stability equation*

$$0 \leq \left(1 - \frac{L}{R_1}\right)\left(1 - \frac{L}{R_2}\right) \leq 1. \tag{4.78}$$

Figure 4.19 is a graph of the *cavity stability equation* and shows that regions underneath the curve in the shaded areas are where stable cavity operation occurs. Using values for the various mirror curvatures as shown in

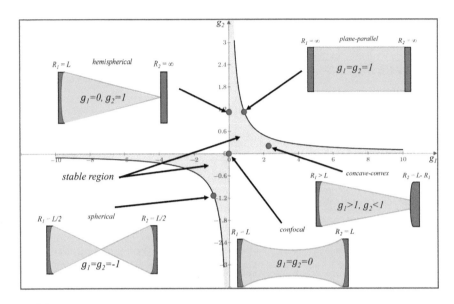

**FIGURE 4.19**
A graph of the *cavity stability equation* showing regions of stability for various configurations.

Figures 4.14–4.18 we can see where the different cavity configurations discussed previously lie on the graph and that they are all stable.

### 4.3.3 Q-factor

The *quality factor*, also often referred to as the Q-factor, and sometimes just as "Q," of an electromagnetic resonator cavity is a dimensionless parameter that is used to describe the damping of an oscillator, bandwidth of the resonator about its center frequency, and the level of losses or gains within the cavity. A resonator with a high Q will have less energy loss and will therefore oscillate longer. The Q of an electromagnetic resonator cavity is typically measured but can be calculated by

$$Q = \frac{2\pi f_o \mathcal{E}}{P} \tag{4.79}$$

where $f_o$ is the resonant frequency of the cavity, $\mathcal{E}$ is the energy stored in the cavity, and $P$ is the power dissipated from the cavity. Realizing that the magnitude of $P$ is

$$P = \frac{d\mathcal{E}}{dt} = \frac{\Delta\mathcal{E}}{\Delta t} \tag{4.80}$$

then Equation 4.79 can be rewritten as

$$Q = \frac{2\pi f_o \varepsilon}{P} = \frac{2\pi f_o \varepsilon \Delta t}{\Delta \varepsilon} = \frac{2\pi f_o \Delta t}{\xi}. \tag{4.81}$$

Here $\xi$ is the cavity loss per oscillation or the energy lost divided by the energy initially stored, and $\Delta t$ is the time for the one oscillation in the cavity. In an optical cavity, $\Delta t$ would be considered to be the round trip time for a photon to travel from one end of the cavity to the other and back again. If the laser cavity has a length, $L$, we can replace $\Delta t$ with $L/c$. We can also replace $f_o$ with $c/\lambda_o$ and Equation 4.81 becomes

$$Q = \frac{2\pi L}{\lambda_o \xi}. \tag{4.82}$$

From the definition that the loss is equal to the energy lost per trip, $\varepsilon_1$, divided by the energy initially stored in the cavity, $\varepsilon_o$, we can find $\xi$ in terms of transmittance, $T$, of the mirrors of the cavity

$$\xi = \frac{\varepsilon_1}{\varepsilon_o} = \frac{nh v T_1 + (1 - T_1) nh v T_2}{nh v} = T_1 + T_2 - T_1 T_2. \tag{4.83}$$

Or using reflectance, $R$,

$$\xi = T_1 + T_2 - T_1 T_2 = 1 - R_1 R_2. \tag{4.84}$$

Equation 4.82 becomes

$$Q = \frac{2\pi L}{\lambda_o (1 - R_1 R_2)}. \tag{4.85}$$

Figure 4.20 is a graph of Equation 4.85 with the product of the reflectance of each mirror varying from 0.001 to approaching 1. The $Q$ of the cavity grows dramatically as the reflectance of each cavity mirror is increased.

Another definition for the $Q$-factor of a laser cavity is the center frequency divided by the bandwidth, or the linewidth divided by the center wavelength. In other words,

$$Q = \frac{f_o}{\Delta f_o} = \frac{\Delta \lambda}{\lambda_o}. \tag{4.86}$$

At high-mirror reflectance Equations 4.85 and 4.86 are approximately equivalent

$$Q = \frac{2\pi L}{\lambda_o (1 - R_1 R_2)} \cong \frac{f_o}{\Delta f_o} = \frac{\Delta \lambda}{\lambda_o}. \tag{4.87}$$

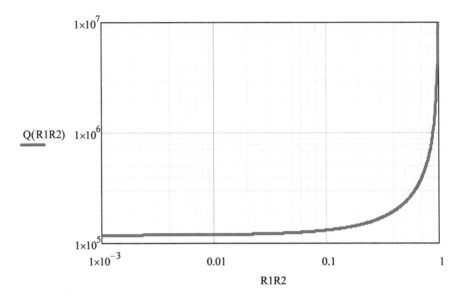

**FIGURE 4.20**
A graph of the Q of a laser cavity as a function of the product of the reflectivity of each mirror.

## 4.4 Chapter Summary

In this chapter, we finally had the tools to understand what a laser actually is. In the first three chapters, we learned about light, amplification, and stimulated emission of radiation. We now have the tools to put it all together and generate laser.

In Section 4.1, we learned of the basic components that make up the laser physically and we developed an understanding of the concepts for *gain, population inversion,* and *saturation.* And then we followed that in Section 4.2 by developing the rate equation to describe what happens during laser operation. With the rate equation we can determine the population of excited and ground-state atoms in the *active medium* and therefore calculate the expected output irradiance for the laser beam.

In Section 4.3, we developed the idea of the *laser cavity, oscillator,* or *resonator.* There are many types of *laser cavity* configurations and we discussed in some detail a few of the more common ones. We also developed a means for calculating the *stability* of the cavity based on the design of the laser mirrors used. Then we developed two definitions for the *quality factor* of the laser cavity and showed how this Q is related to both the reflectance of the mirrors and the linewidth of the laser.

## 4.5 Questions and Problems

1. What is an *active gain medium*?
2. What is the chemical symbol for neodymium-doped yttrium orthovanadate?
3. What is an *output coupler*?
4. What is the difference between a *reflector* and an *output coupler*?
5. List and describe the four basic components of the laser.
6. Use Equation 4.9 to calculate the irradiance through a cavity of 4, 5, and 10 passes? Use the real numbers from a typical $Nd^{+3}$:$YVO_4$ laser pointer-type active medium, where $L = 0.01\,cm$, $G = 315\,cm^{-1}$, $R_c = 0.5$, and $I_0 = 1\,\mu W/cm^2$.
7. What happens as n is increased from 4 to 10 in Problem 6?
8. What is the *cross section coefficient for absorption*?
9. How is the *cross section coefficient for absorption* related to the *gain* of the laser?
10. What is the *saturation* limit?
11. For a two-level laser medium what is meant by $\dfrac{N_2}{N_1} < 1$?
12. For a two-level laser medium what is meant by $\dfrac{N_2}{N_1} > 1$?
13. What is the *quantum defect* of the laser?
14. What is *spatial hole burning*?
15. What is *spectral hole burning*?
16. In our discussion of the four-level laser, why did we not bother with rate equations for the states $n_4$ and $n_1$?
17. Draw a four-level laser energy level diagram and label all coefficients.
18. Write the rate equations for all four levels of Problem 17.
19. Write the rate equation for the photon number within the laser cavity for the system described in Problems 17 and 18.
20. Given a 20% transmittance *output coupler* for a laser has a diameter of 1 cm, the length of the cavity is $L = 10\,cm$, the length of the gain medium is $l = 7\,cm$, the wavelength of the laser is 532 nm, the index of refraction of the gain medium is $n = 1.2$, and the output irradiance is $10\,mW/cm^2$. What is the number of photons in the cavity? (Hint: use Equation 4.62)
21. What is the *stability parameter* of a laser cavity?
22. Given a cavity of length $L = 1\,m$ and radius of curvature of one mirror is $\infty$ and the other is $5\,m$. Is this cavity stable?

23. Given the reflectance of both mirrors in a cavity is 99% and the cavity is 1 m long. What is the linewidth of this cavity?

24. If the center wavelength of Problem 23 is 532 nm what is the $Q$ of the cavity?

25. What is the frequency bandwidth of the system in Problems 23 and 24?

# 5

## *What Are Some Types of Lasers?*

As we discussed in Chapter 1 the word laser is an acronym that actually stands for Light Amplification by the Stimulated Emission of Radiation. We started by looking at the acronym itself as the simplest method to determine where to start in our study of the laser. The very first word in the acronym is "light." So, we discussed in great detail in Chapter 1 the answer to the question: "What is light?"

It makes sense that our next discussion, as given in Chapter 2, was the next major word in the acronym itself which is "amplification." We discussed amplification and amplifiers in some detail. Chapter 3 then followed with the question: What is "stimulated emission of radiation?" Finally, we developed all the tools necessary to discuss the concept of the laser. In Chapter 4, we went into detail about how the laser process actually works and the underlying physics and math to describe it. Understanding the physics and math is one thing, but to truly become a laser scientist or engineer we also need to understand, in a pragmatic way, how lasers are actually physically put together.

There are many types of lasers such as solid-state, gas, liquid, chemical, semiconductor, metal-vapor, and there are even exotic systems requiring nuclear detonations or antimatter annihilation. The lasers available today range in power output levels from nearly single photon sources to megawatt capability with more photons than countable. In this chapter, we will discuss some of these types of lasers and their components. As we begin to see how various lasers are realized we will begin to be able to think of how they might be useful for the many applications they are used for today.

## 5.1 Solid-State Lasers

### 5.1.1 Ruby Laser

The ruby laser was the first ever laser to be constructed. It was built by Theodore H. Maiman at Hughes Research Laboratories on May 16, 1960. The laser uses a synthetic ruby crystal rod as the active gain medium and produces milliseconds long pulses of red light at a wavelength of 694.3nm. Figure 5.1 shows a typical arrangement for a ruby laser.

As shown in Figure 5.1 the ruby laser uses electrical power, which is then converted into optical power by flashlamps that generate bright broadband flashes of light much like a flash on a camera. The broadband flashes are then absorbed by the synthetic ruby rod causing a population inversion. The cavity is arranged with mirrors at each end of the rod with a highly reflective mirror on one end and an output coupler at the other. Figure 5.2 shows the "laser head" of an actual ruby laser construction. The laser has two linear flashlamps and the ruby rod active medium, which are all placed inside a reflective cavity in order to make use of as much of the optical pump pulse as possible. The ruby rod contains red chromium ions, which provides the stimulated emission transitions for laser action.

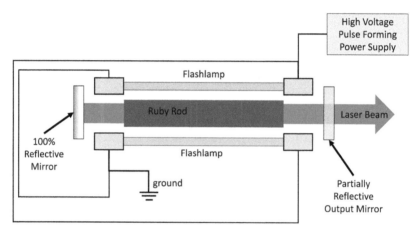

**FIGURE 5.1**
The typical ruby laser.

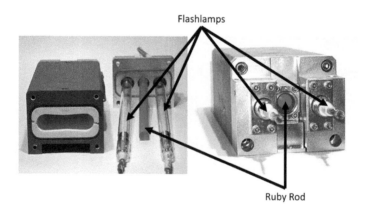

**FIGURE 5.2**
A typical ruby laser configuration has two linear flashlamps with the ruby rod in the middle all surrounded by a reflector system.

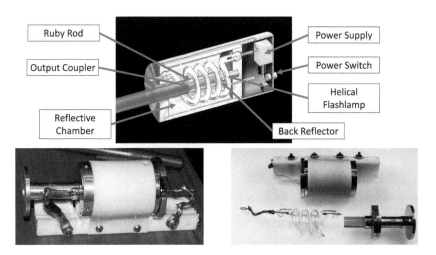

**FIGURE 5.3**
The first ever working laser was the ruby laser built by T. H. Maiman in 1960 using a helical flashlamp and a synthetic ruby rod.

Figure 5.3 shows the original ruby laser built by Maiman, which is slightly different from the typical device shown in Figures 5.1 and 5.2. While the flashlamp was still "linear" in that the discharge pulse traveled from one electrode at one end to the other electrode at the other end, the flashlamp itself was constructed in such a way as to wrap around the active medium ruby rod in order to enhance the coupling of pump light into the active medium. This type of flashlamp is referred to as a "helical flashlamp."

### 5.1.2 The Neodymium-Doped Yttrium Aluminum Garnet Laser

Yttrium aluminum garnet or YAG is a crystal that is doped with ionized neodymium. It is the neodymium ions that provide the stimulated emission transitions for laser action. This laser is often referred to as a Nd:YAG, and oftentimes just "YAG," or more concisely it is $Nd^+{:}Y_3Al_5O_{12}$. Figure 5.4 shows a diagram of a Nd:YAG laser. The function is much the same as the ruby laser only the medium has different transition energies. The Nd:YAG typically is implemented in a cavity that selects the 1,064-nm wavelength transition, which is in the infrared and cannot be seen by the human eye. There are other transitions sometimes used at 946, 1,120, 1,320, and 1,440 nm.

Nd:YAG lasers can be flashlamp, diode, or laser pumped and can be continuous or pulsed. In the pulsed mode, they are typically operated by pumping the laser until an extreme population inversion is reached and then suddenly the output pathway for transmission of light is opened allowing the majority of cavity-stored energy to be dumped out at once. This process is known as *Q-switching* and is usually accomplished with some sort of nonlinear optical switching component.

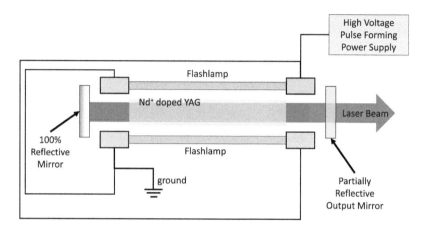

**FIGURE 5.4**
The Nd:YAG laser is very much the same as the ruby laser but with a different active medium and outputs infrared light.

In this laser, the YAG is the "host material" for the $Nd^+$ "dopant" material. There are other host materials that will work as well for neodymium such as yttrium orthovanadate or $YVO_4$ as discussed in Chapter 4, yttrium lithium fluoride, or YLF (pronounced "yelf"), and even glass. All of these lase in the infrared. We are most likely familiar with these lasers as we have seen very bright green beams at laser light shows, or bright green laser pointers. But these are infrared lasers so how are they emitting green beams of light?

### 5.1.2.1 Second-Harmonic Generation

The infrared beams are made to be incident on a special nonlinear crystal known as a "frequency doubling crystal" or a "second harmonic generation crystal" as shown in Figure 5.5. An incident number of photons, $n$, at an infrared wavelength goes into the crystal and through this nonlinear process

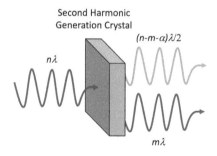

**FIGURE 5.5**
A second-harmonic generation crystal is used to convert infrared photons into green photons in the Nd:YAG laser.

the frequency is doubled (wavelength is halved). The process has some loss, $\alpha$, and some number of infrared photons, $m$, that do not get converted and pass through unencumbered. The number of photons at the new frequency, $n$-$m$-$\alpha$, exits the crystal.

So, the 1,064-nm infrared photons enter the crystal and 532-nm photons exit it. There are some of the infrared photons that do not get converted in the process and so a filter is usually placed after the crystal to block the infrared and pass the green light. Also, the process is a threshold process so there is a minimum amount of laser energy required to trigger the second-harmonic generation effect. The crystals themselves are made of many different materials but the most common process for converting near-infrared beams to visible is potassium titanyl phosphate which has the chemical composition of $KTiOPO_4$ and is abbreviated as KTP. Figure 5.6 shows the complexity of the crystal's structure. It should be noted here that KTP is easily damaged at high powers. Other materials such as lithium triborate (LBO) and beta barium borate (BBO) are used when high power is needed as they are more robust crystals.

There are many host materials and dopant materials for solid-state lasers and the pumping mechanisms can be from flashlamps, other lasers,

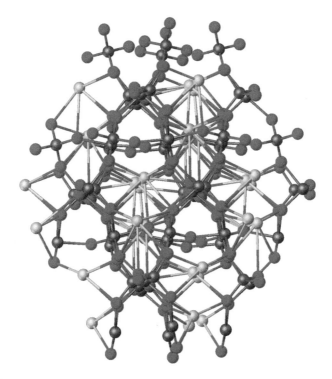

**FIGURE 5.6**
The complex KTP crystal is a commonly used "doubling crystal" in the green laser pointer. (This file is licensed under the Creative Commons Attribution-Share Alike 4.0 International license and was produced by the author Smokefoot.)

light-emitting diodes, electric arcs, and even focused sunlight. The combinations are nearly endless and new solid-state laser configurations and materials combinations are invented all the time.

### 5.1.3 The Diode-Pumped Solid-State Laser

In Section 4.1 the $Nd^{+3}$:$YVO_4$ laser active medium was discussed. This laser is the typical green laser pointer that is sold commercially that most people have seen whether they realize it or not. The laser uses a red diode laser at 808-nm wavelength to pump the neodymium ions doped in the orthovanadate host material. In return 1,064-nm wavelength light is produced.

Figure 5.7 shows a diagram of the $Nd^{+3}$:$YVO_4$ laser. The laser is actually two lasers in one. The pump source being a diode laser (to be discussed later in more detail), which pumps the active medium. The cavity mirrors are typically glued directly to the gain medium on each side. The back reflector mirror is a specially coated mirror that allows the 808-nm pump beam through but reflects the 1,064-nm light. The output coupler is partially reflective at 1,064 nm and in some cases is between the orthovanadate crystal and the "doubling crystal." In some cases, the "doubling crystal" is glued to the orthovanadate crystal and the output coupler is then glued to the outside of the "doubling crystal." This is a design choice typically and varies from laser to laser. There are then usually some beam shaping and collimating optics and a final filter that blocks any 1,064-nm infrared light that makes it through the crystal and other optics. The filter passes the green 532-nm laser beam.

### 5.1.4 The Fiber Laser

A fiber laser is a type of solid-state laser that uses specially doped fiber optic cable as the active gain medium. The fiber is typically doped with rare-earth

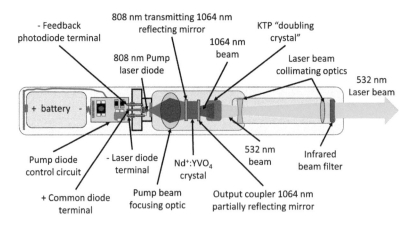

**FIGURE 5.7**
The diode-pumped $Nd^{+3}$:$YVO_4$ laser is a commonly used green laser pointer.

elements such as erbium, ytterbium, neodymium, dysprosium, praseodymium, thulium, and holmium. These lasers are much like the Nd:YAG solid-state laser only instead of a host rod material there is a host fiber material. Also, the fiber is typically "end-pumped" with diode lasers that emit in the active fiber material's absorption band for singlet-state transitions.

Figure 5.8 shows a typical configuration for a fiber laser. Multiple diode lasers provide pump light into the larger and much longer spool of active fiber. There is typically some type of output coupling and collimating optical system often referred to as the "laser head" where the beam exits. Fiber lasers are very scalable in power output. More diode pump lasers can be added and longer or parallel active fiber spools can be added. Multiple active fiber spools can then be combined into one larger fiber output system. Many kilowatts of output power can be achieved with these types of lasers. In fact, these types of lasers are used for industrial laser welding systems and experimental laser weapon systems. Figure 5.9 shows a commercial 10 kW laser welding system sold by IPG Photonics.

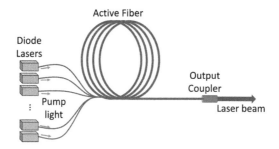

**FIGURE 5.8**
The fiber laser can be scaled up to very high-output power.

**FIGURE 5.9**
The 10 kW fiber laser from IPG Photonics.

## 5.2 Dye Lasers

### 5.2.1 The Flashlamp-Pumped Liquid Dye Laser

One of the most versatile lasers available is the flashlamp-pumped liquid dye laser. This type of laser converts electrical energy stored in capacitors to an optical flashlamp pulse, which is then used to pump the singlet transition of organic molecules in various dyes dissolved in a liquid solvent host. The dye is typically a very complex (an oftentimes very toxic) powder and the solvent host is most often methanol, ethanol, water, or a mixture of these solvents. Sometimes the solvent is more exotic depending on the dye.

The most common active medium used in these lasers is Rhodamine 6G, sometimes called Rhodamine 590 Chloride, dissolved in methanol or ethanol. Figure 5.10 shows the Rhodamine 6G molecule. The molecule is quite complex and has many electrons in various orbitals that can be excited to singlet states as well as triplet states. Therefore, the molecule has a very broadband absorption curve (meaning it can absorb light at many wavelengths) and a very broadband fluorescence curve (meaning it can emit light at many wavelengths). This makes the dye molecule suitable as a laser medium when either broadband or tunability is needed for the laser application.

Figure 5.11 shows a typical configuration for a linear flashlamp-pumped liquid dye laser. A high-voltage pulse is input to the flashlamps where they in turn flash. The broadband flashlamp pulse is absorbed by the liquid dye flowing through the laser head. A highly reflective mirror and a partially reflective output coupler create the laser cavity and these mirrors can be coated for broadband lasing or narrow band lasing which will help select the transitions desired.

We should point out here a design requirement for our pump source. In Chapter 3, Example 3.1, we discussed triplet states and how a population inversion can be trapped in a triplet state with long decay lifetimes. Many organic dyes have such triplet states available and in order to achieve a population inversion the optical pump pulse must be short enough as to not allow for "triplet state trapping." For Rhodamine 6G in methanol an optical

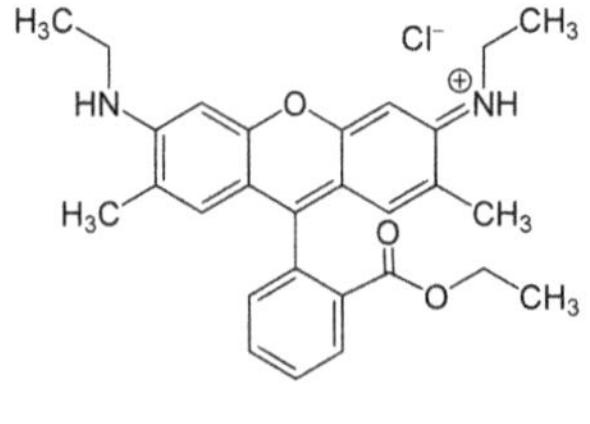

**FIGURE 5.10**
The Rhodamine 6G laser dye molecule.

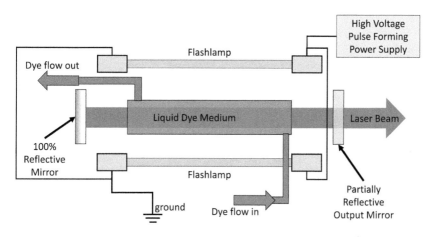

**FIGURE 5.11**
The typical configuration for a linear flashlamp-pumped dye laser.

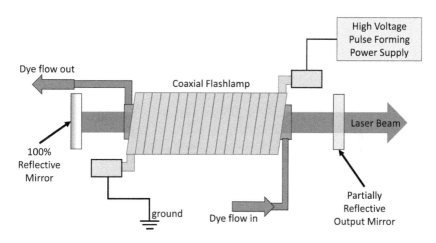

**FIGURE 5.12**
The coaxial flashlamp-pumped dye laser.

flashlamp pulse must be on the order of a couple of microseconds or shorter or this effect will become an issue.

Many flashlamp-pumped liquid dye lasers make use of a special type of flashlamp called a *coaxial* flashlamp. Figure 5.12 shows a diagram of the *coaxial* flashlamp and Figure 5.13 shows an actual one on a ruggedized dye laser system built by the U.S. Army in Huntsville, Al. The lamp itself is constructed by placing a smaller cylindrical glass tube inside a larger one of same length. The tubes are then filled with xenon gas and sealed on each end with an electrode. When the electrical energy is dumped into the lamp the arc within the xenon gas forms in a cylindrical sheet from one end to the other coaxially.

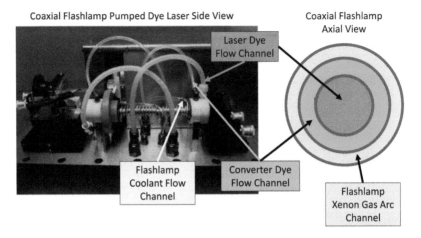

**FIGURE 5.13**
The coaxial flashlamp-pumped dye laser with converter dye channel. (Picture courtesy of the U.S. Army.)

Figure 5.13 shows another unique aspect the coaxial configuration enables with liquid dye lasers. If a small enough tube for the laser dye is used in the middle of the flashlamp then there will be an empty space between the dye-active medium tube and the flashlamp's inner tube. This empty space is typically used for flowing cooling water or other transparent liquids through the system to cool the flashlamp during long duration cycles of many pulses. Since flashlamp outputs are so broadband across the visible they can be inefficient in delivering light at the right absorption wavelength to the active medium (dye in this case). With the coaxial flashlamp configuration shown in Figure 5.13 a clever trick can be used to increase the efficiency of the conversion of pump light to laser light. A dye that absorbs the flashlamp light that isn't absorbed by the laser dye is added to the cooling loop. For Rhodamine 6G which absorbs in the green to yellow and lases from yellow to red all of the violet to green light emitted by the flashlamp is wasted, so a dye that absorbs in that region and emits in the green to yellow is best suited to be added between the flashlamp and the laser dye. This type of dye is called a "converter dye" because it converts the unused pump light to useable pump light.

Figure 5.14 shows graphs of the output irradiance of a typical xenon flashlamp as a function of wavelength overlaid on the absorption and fluorescence curves of Rhodamine 6G laser dye and of a typical converter dye. The graph shows how implementing a converter dye will enable more of the flashlamp output light to be used to pump the laser dye.

### 5.2.2 Laser-Pumped Liquid Dye Lasers

The most efficient way to be certain that almost every pump photon generated is converted into laser light is to pump with a single wavelength

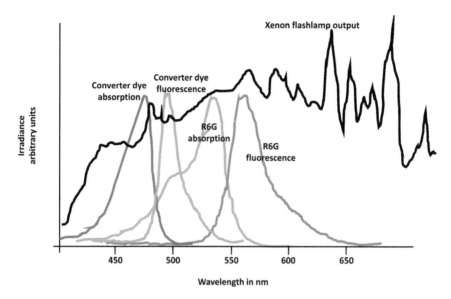

**FIGURE 5.14**
The output spectra of a xenon flashlamp overlaid on the absorption and fluorescence curves of Rhodamine 6G and converter dyes.

that matches the active medium's singlet-state transition absorption. In other words, we pump the medium at the peak of its absorption curve with a laser of that wavelength or as near to it as possible. We have already discussed the diode laser-pumped solid-state laser. The laser-pumped liquid dye laser is very similar and, in some cases can even be pumped by diode lasers.

Figure 5.15 shows a typical configuration for laser pumping a liquid dye cell. For low-power systems the dye cell is typically just a quartz cuvette filled with the liquid dye. The cavity is typically made of two flat mirrors: one near 100% reflective mirror and one partially reflective one. The pump laser beam is incident on the cuvette orthogonally. Sometimes some pump laser beam shaping optics are required to focus the beam on the cuvette in the right place with enough photons to generate the population inversion. For the configuration shown in Figure 5.15 if the laser dye is Rhodamine 6G the typical pump laser is a pulsed and frequency-doubled Nd:YAG. The 532-nm output of the "doubled YAG" is very near to the peak of the absorption curve for Rhodamine 6G and is very efficient at pumping it without wasting photons.

For other wavelengths from other laser dyes other pump lasers at different wavelengths are used. Another example is when using green or blue laser dyes typically known as Coumarin dyes. Coumarin 480 is a dye that lases very blue. For low-power purposes in the configuration shown in Figure 5.15 a nitrogen gas laser (to be discussed later) is an ideal pump laser as it lases at a wavelength of 337.1 nm which is in the absorption curve or "pump band" of

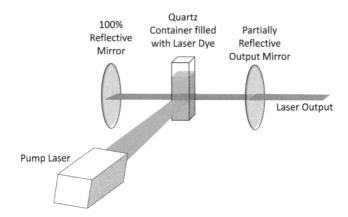

**FIGURE 5.15**
The typical configuration for laser-pumping liquid dye lasers.

most of the blue to green dyes. In fact, the absorption curve for Rhodamine 6G is so broad that it will lase when pumped with the nitrogen laser but the conversion of pump light to laser light is less efficient than pumping it closer to the peak of the dye's absorption curve.

### 5.2.3 Solid-State Dye Lasers

Just like solid-state crystal or glass-based laser, the dye molecule can be doped into various polymers and crystal materials. One of the most common materials that laser dyes are doped into is poly (methyl methacrylate) or PMMA. One of the most commercially common applications of doping these plastics with laser dyes is in plastic protractors and clipboards as shown in Figure 5.16. In fact, these cheap protractors and clipboards can actually be used as the active medium for a laser-pumped solid-state dye laser.

**FIGURE 5.16**
Laser dyes can be doped into modern plastic materials that can be used for laser-active media. (Note: objects are not to scale.)

During the 1990s, the U.S. Army in Huntsville, Al did extensive research on doping various solid host materials with various laser dyes with significant success. The solid hosts are quite suitable for laser pumping but do not hold up to the thermal issues involved with multiple flashlamp pulses. Dye-doped plastics will lase when flashlamp pumped until too much heat builds up inside the material causing random index of refraction changes throughout the material that function much like atmospheric turbulence. These randomly distributed index of refraction changes throughout the solid host material act as randomly placed lenses of all types and focus lengths, therefore, disrupting the optical laser cavity pathway. The best solution to the issue was to laser pump directly into the absorption band of the dye and the thermal problems went away.

### 5.2.4 Continuous Wave Output Dye Lasers

Implementing dye lasers in a continuous wave mode with the beam being continuous requires a bit of engineering based on the physics of the laser dye's stimulated emission process. As was discussed previously, the organic dye molecules have many available electronic states and can get trapped into a triplet state that will not lase. Continuously pumping the same volume of dye mixture for more than a microsecond or so will typically lead to triplet-state excitation, singlet-state transitions will be shut out, and the laser action stopped.

Figure 5.17 shows how to avoid this triplet-state problem by moving the volume of dye being pumped by the pump laser to be flowed out of the system before it can be trapped in the triplet state. This means that the entire lasing volume must be moved out of the pump beam within a microsecond or faster. The flow speed of the dye within the cell is very fast, on the order

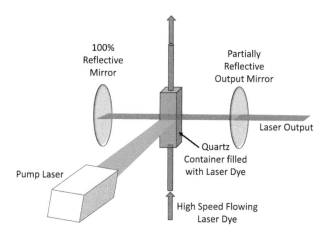

100%
Reflective
Mirror

Partially
Reflective
Output Mirror

Laser Output

Quartz
Container filled
with Laser Dye

Pump Laser

High Speed Flowing
Laser Dye

**FIGURE 5.17**
The continuous wave dye laser requires a very fast flow loop.

of 1,000 m/s! While Figure 5.17 shows flow through a cell, many modern continuous wave dye lasers actually flow the dye mixture through a jet nozzle in order to achieve the speeds required. The engineering details of this type of laser are quite involved and construction materials for the optical system require rigid mechanical stabilization due to vibrations in the flow loops.

## 5.3  Semiconductor Lasers

### 5.3.1  Diode Lasers

A laser diode is a modern solid-state electronics device made of semiconductor materials much in the same way as a light-emitting diode or LED. In the diode laser, the laser beam is generated at the diode junction between two different types of semiconductor materials.

Figure 5.18 shows a diagram of the basic diode. Two materials, a p-type and an n-type, are placed in intimate contact with each other. The p-type material has the ability to supply positively charged "holes." In other words, the material can receive negatively charged electrons. The n-type material can supply negatively charged electrons. When a voltage source is connected as shown in Figure 5.18, where the positive side is connected to the p-type material and the negative lead is connected to the n-type material this places the diode junction in what is known as "forward bias." While in forward bias at the region near the junction, called the "depletion region" electrons will gather on the p-type side of the junction and holes will gather at the

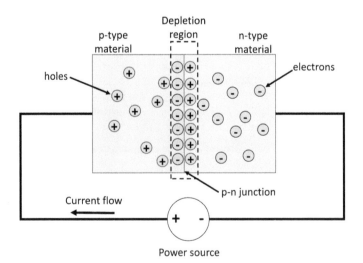

**FIGURE 5.18**
The basic diode configuration.

Typical Laser Diode

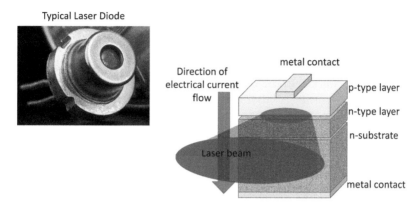

**FIGURE 5.19**
The typical laser diode configuration.

n-type side of the junction. It is when these electrons and holes recombine that a photon is generated. If the materials are doped right then a population inversion can be created in the forward bias and therefore lasing can occur.

Figure 5.19 shows a typical diode laser and how the materials are configured within it. Note that for the standard diode laser the laser axis is orthogonal to the flow of electrical current. Typical materials used to construct diode lasers are gallium arsenide (GaAs) and aluminum gallium arsenide (AlGaAs). There are also many different layering techniques and cavity configurations for the laser cavities orthogonal to current flow direction.

### 5.3.2 Vertical-Cavity Surface-Emitting Laser

Figure 5.20 shows a different configuration for a diode laser. This type of laser has the laser cavity along the same axis as current flow and light

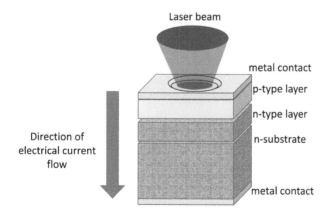

**FIGURE 5.20**
The VCSEL along the same axis as the electrical current flow direction.

is emitted from the surface of the top layer of semiconductor material. It is called a vertical-cavity surface-emitting laser or VCSEL (pronounced "vic sull").

This type of diode laser enables a freedom of manufacturing choices. Many VCSELs can be produced in an array side by side on a single wafer. This allows for creating a large array of surface emitting lasers that can be addressed all at once or individually.

## 5.4 Gas Lasers

### 5.4.1 The Helium–Neon Laser

The helium–neon laser or HeNe laser and sometimes just "HeNe" pronounced "he knee", is one of the most commonly used lasers in laboratories across the world. The well-known continuous-wave red laser emits a wavelength of 632.8 nm with a very narrow linewidth and can be made very stable in both spatial and temporal output modes. The HeNe was first demonstrated emitting an infrared wavelength in 1960 and then 2 years later in 1962 at Bell Telephone Laboratories the red line was demonstrated. The HeNe is available in modern day in multiple output wavelengths in the green, yellow, orange, red, and infrared since the gas mixture of helium and neon has so many available singlet transitions.

Figure 5.21 shows a HeNe with all of the components labeled. The laser's gas reservoir is filled with a mixture of helium and neon in a 10:1 ratio at pressures lower than atmospheric pressure. Electrodes at each end of the gas-filled tube are connected to a high-voltage power source that supplies

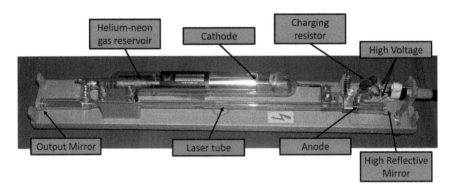

**FIGURE 5.21**
The component of the HeNe laser. (The picture is a modified GNU Free Documentation License Version 1.2 image published by the Free Software Foundation.)

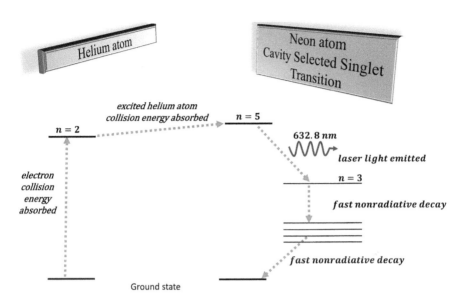

**FIGURE 5.22**
The selected transition of the red HeNe laser.

energy to the gas, which in turn creates a plasma discharge within the tube. Figure 5.22 shows the laser transition process for the HeNe 632.8-nm output. Electrons flowing through the plasma from one electrode to the other collide with and excite the helium atoms to an excited state in the $n = 2$ orbital. The excited helium atoms then collide with neon atoms transferring the energy to them. The neon atoms are excited to the $n = 5$ orbital and s suborbital or 5s. The electron transitions from the 5s to the 3p level by emitting a photon at 632.8 nm. The neon atom then rapidly decays back to the ground state. Note that optical coatings on the mirrors are used to select the specific 632.8-nm transition.

We should point out here that the HeNe is most likely the most used laser by the laser scientists and engineers around the globe. They are used as stable photon sources for measuring distances and times, for aligning other optical systems and lasers, and are the "work horse" of the laser laboratory.

## 5.4.2 The Carbon Dioxide Laser

The carbon dioxide or $CO_2$ laser was invented by Kumar Patel at Bell Laboratories in 1964 and is widely used for industrial and medical applications today. The $CO_2$ laser is much like the HeNe laser in that it uses a gas mixture that relies on collisional energy transfer and it is a continuous wave output. The $CO_2$ laser can be scaled to very high-output power and has lasing wavelengths in the infrared at 9.4 and 10.6 μm.

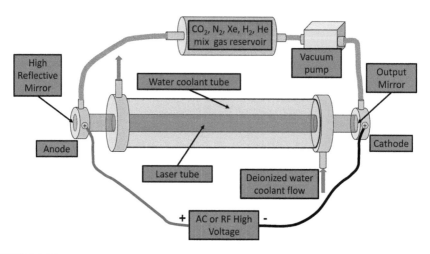

**FIGURE 5.23**
The carbon dioxide laser can be scaled up to very high-continuous output power levels.

Figure 5.23 shows a diagram of the $CO_2$ laser. The gas reservoir is filled with carbon dioxide, nitrogen ($N_2$), hydrogen ($H_2$), xenon (Xe), and helium (He) gases. The mixture ratios vary but about 50%–60% of the gas is He with $CO_2$ and $N_2$ each about 10%–20%. The other gases are single-digit percentages or even less in the mixture. Electrons flowing through the gas tube create a plasma discharge just like in the HeNe.

In the case of the $CO_2$ laser, the electrons collide with the $N_2$ molecule exciting it. In turn, the excited $N_2$ molecule collides with the $CO_2$ molecule exciting it. The excited $CO_2$ molecule provides the laser transition. The $N_2$ molecule colliding with the $CO_2$ only reduces the $N_2$ to a lower-excited state. It must, therefore, collide with a He atom to release the remaining energy and return to ground state. The He atom is heated up and transfers energy to the walls of the system through vibrational collisions. In fact, a significant amount of heat is generated during this laser process so the $CO_2$ laser must be cooled either by flowing air or water depending on the power output level of the laser.

Figure 5.24 shows a 15 kW $CO_2$ laser while in operation at a U.S. Air Force laboratory. The glow discharge is very clearly seen in the image. While the beam is far in the infrared and cannot be seen by the human eye, the results of the beam can. The target block on the right of the picture is in flames due to the incident laser beam.

### 5.4.3 The Nitrogen Laser

The nitrogen laser is a transversely excited atmospheric (TEA) laser that operates in the ultraviolet at 337.1 nm. It is likely one of the most commonly "homebuilt" or "do-it-yourself" laser due to an article published in *Scientific*

**FIGURE 5.24**
A 15 kW $CO_2$ laser burning a block target. (Courtesy U.S. Air Force.)

*American* in 1974, explaining how to build one with amateur construction techniques and materials from a local hardware store or hobby shop.

The basic construction design for the $N_2$ laser is shown in Figure 5.25. A large capacitor is constructed using a dielectric board with a conductive surface (typically copper) on one side and two conductive regions on the other side. A gap between the two conductive plates on the top side is where a discharge gap is created within a chamber filled with $N_2$ gas. As the capacitor charges up to a voltage high enough to break down the spark gap, it then discharges across the discharge channel. The electrons flowing across the channel gap excite the $N_2$ gas molecules. The excited molecules emit 337.1-nm light along the laser path. The gain of the laser is so high that it will lase without mirrors. This type of laser is called "super-radiant" and

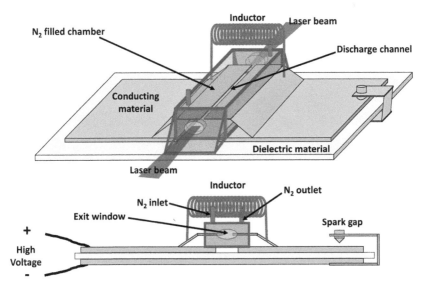

**FIGURE 5.25**
The nitrogen laser can be constructed very easily with hardware store-type materials.

sometimes "superluminescent." Adding mirrors increases the gain and laser beam output.

The nitrogen laser has very short pulses ranging from picoseconds to nanoseconds. It is low-energy per pulse on the order of 1 μJ to 1 mJ. However, it can have a very high-repetition rate in the kHz range. So, the $N_2$ laser can have a fairly high-peak output power even into the megawatts.

### 5.4.4 The Gas Dynamic Laser

The gas lasers we have discussed thus far have consisted of systems that were filled with various gas mixtures and the active media were for the most part static in flow speed (or no flow at all) compared to the laser action. Another type of gas laser, known as the gas dynamic laser, makes use of difference between relaxation times between different excited states. The gas-active medium will be chosen such that two excited states can be achieved. One of the states will relax faster than the other. The gas is flowed at a specific velocity through the laser cavity of a specific length so that the faster decaying state has relaxed but the slower one has not. This creates the population inversion needed for laser action.

Figure 5.26 shows a typical configuration for a gas dynamic laser. The device is practically a rocket engine where fuel is flowed into a combustion chamber where it is burned and heated and forced through nozzles. The flow through the nozzles increases the speed to as high as supersonic rates. Once the gas is sped up through the nozzles and is flowed down a chamber for the appropriate distance the population inversion is formed between the optical components of the laser cavity. The gas is then flowed through an exhaust system and is recovered or disposed of.

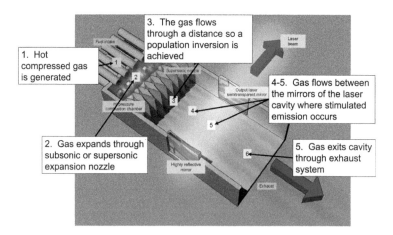

**FIGURE 5.26**
The gas dynamic laser uses rocket engine-type flow to accelerate two active media species with different relaxation times to create the population inversion.

### 5.4.5 The Excimer Laser

When certain noble gas atoms (argon, krypton, or xenon) combine with reactive gasses (fluorine or chlorine) a "pseudo-molecule" is formed in an excited state. This excited molecule is called an "excited dimer" or "excimer" for short. Another term sometimes used is "excited complex" or "exciplex." The excimer laser is sometimes referred to as an exciplex laser. In fact, there is a bit of terminology sloppiness here in that an excimer molecule can only be formed with excited states of the noble gas atoms with themselves. Exciplex molecules are formed between the excited noble gas atom and a different reactive gas.

The excimer molecules are generated by exciting the noble gas atoms with an electrical discharge or with high-energy electron beams. Once the noble gas atoms are excited, they can form temporary exciplex molecules. If these molecules are in an optical cavity stimulated emission can occur and a laser beam is formed. Once the molecules de-excite the atoms dissociate from each other.

The wavelength of the excimer or exciplex laser is dependent upon the molecules formed. One of the most common exciplex lasers is the xenon-chloride (XeCl) laser, which operates at 308 nm in the ultraviolet. Both excimer and exciplex lasers are typically ultraviolet lasers. The most common use for these types of lasers is with photolithography and micro-electronic chip manufacturing. They are also used for medical applications such as angioplasty and refractive eye surgery.

---

## 5.5 Chemical Lasers

### 5.5.1 The Hydrogen Fluoride Laser

Chemical lasers use chemical reactions to generate an excited-state medium population inversion. The hydrogen fluoride (HF) laser was first demonstrated as a continuous wave laser operating in the infrared at about 2.8 μm wavelength in 1972 by The Aerospace Corporation in El Segundo, California. The patent disclosure describes how the laser operates as follows:

> In this invention … the requisite population inversion of the active gaseous medium is provided by a chemical reaction involving the diffusion of a first reactant material into a high-speed flow containing a second reactant material. The two reactant materials react chemically to provide a sustained flow of a vibrationally excited molecular gas capable of lasing.
>
> The laser is chemically pumped because the vibrational population inversion and radiation energy is provided directly by a chemical reaction.

> The technique of this invention employs a continuous high-speed flow system which provides freshly activated species in the optical cavity and at the same time provides for rapid removal of spent molecules.

Figure 5.27 shows a modern-day configuration for an HF laser. The laser generates fluorine by combusting nitrogen trifluoride ($NF_3$) with ethylene ($C_2H_4$) or deuterium ($D_2$). Burning these chemicals together not only generates fluorine but also heat and therefore pressure for the flow system. At this point, the fluorine gas is fed into nozzles simultaneously with hydrogen ($H_2$) gas. The $H_2$ and F react yielding a reaction product of excited HF in the optical cavity. The flow rates are specifically designed much like the gas dynamic laser so that the population inversion is maintained between the optical cavity components and then they are flowed out through an exhaust and recovery system.

### 5.5.2 The Deuterium Fluoride Laser

While the HF laser is quite scalable, the lasing wavelength at 2.8 μm does not propagate through the atmosphere very well. HF lasers are best suitable for vacuum propagation applications which imply space and high cost. Fortunately, the same laser system components and concept works well by simply changing out the hydrogen with deuterium and doing so shifts the lasing wavelength to 3.8 μm, which propagates through the atmosphere very well.

Figure 5.28 shows the deuterium fluoride laser, which is much the same as the HF laser. The major difference is that instead of injecting hydrogen at the nozzles a mixture of deuterium and helium is injected. Also, the optics are coated for 3.8 μm operation rather than the 2.8 μm.

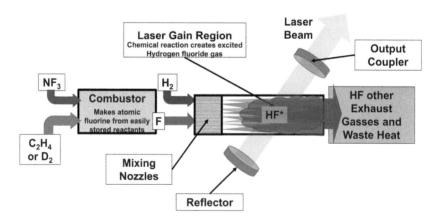

**FIGURE 5.27**
The HF laser reacts hydrogen and fluorine to create excited HF.

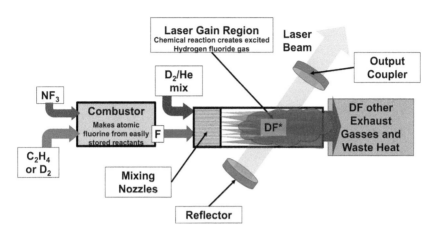

**FIGURE 5.28**
The deuterium fluoride (DF) laser reacts deuterium and fluorine to create excited DF. The DF laser beam propagates well in the atmosphere.

### 5.5.3 The Chemical Oxygen Iodine Laser

Another chemical laser is the chemical oxygen iodine laser or COIL. As shown in Figure 5.29, this laser uses a mixture of chlorine gas (Cl), molecular iodine (I), and an aqueous mixture of hydrogen peroxide ($H_2O_2$) and potassium hydroxide (KOH). The aqueous mixture is reacted with the chlorine gas that yields reaction products of potassium chloride salt, heat, and oxygen in an excited state. Much like with the HeNe laser the excited oxygen is then mixed with molecular iodine where the energy is transferred

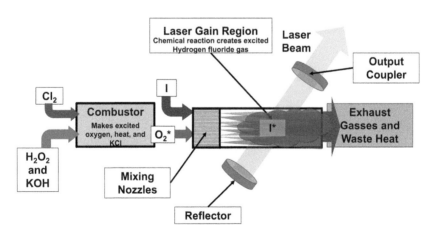

**FIGURE 5.29**
The COIL reacts with chemicals to produce excited oxygen, which is then collided with iodine to generate stimulated emission.

from the oxygen to the iodine via collisions. A population inversion of excited iodine then emits laser light at 1.315 μm.

## 5.6 Metal Vapor Lasers

### 5.6.1 The Copper Vapor Laser

Figure 5.30 shows the configuration for a copper vapor laser. A ceramic tube houses copper pellets and a low-pressure buffer gas like neon. A pulsed electrical discharge is set off between electrodes, which increases the temperature within the tube to over 1,450°C. The high temperature produces copper vapor at low pressures. Electrons within the discharge then collide with the copper vapor atoms exciting them to an upper-state population inversion. The laser transition wavelength can be selected between two transitions: green at 510 nm and yellow at 578 nm. The laser self-quenches as lower-level excited states take longer to relax than the laser transitions. The pulses typically are 5–60 ns in duration.

Copper vapor lasers are capable of producing very high-power output in the megawatts with repetition rates of as much as 100 kHz. They are often used for machining purposes and for laser light shows. The lasers also are known for being very stable in output irradiance and beam quality.

### 5.6.2 The Helium–Cadmium Laser

Another metal vapor laser similar to the copper vapor laser is the helium–cadmium (HeCd) laser. Typical HeCd lasers can be selected for transitions at a blue 442 nm or violet 325 nm wavelength output. There are other

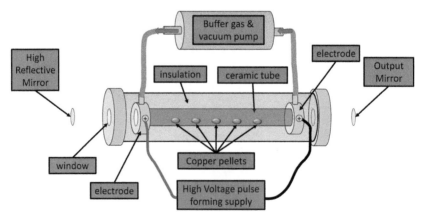

**FIGURE 5.30**
The copper vapor laser.

transitions with this laser as well with a possible of up to 12 different output wavelengths ranging from red to violet.

### 5.6.3 Other Metal Vapor Lasers

There are many metal vapor lasers available that operate similarly to the ones discussed thus far. Some of these include helium–mercury (HeHg), helium–selenium (HeSe), and helium–silver (HeAg). There are strontium and manganese vapor lasers as well.

## 5.7 Ion Lasers

### 5.7.1 The Argon Ion Laser

Invented in 1964 by William Bridges at the Hughes Aircraft Company, the argon (Ar) ion laser uses ionized argon gas generated from an electrical discharge as the active medium. The typical $Ar^+$ laser uses a ceramic tube filled with a high-density plasma of argon ions. In some cases, a magnetic field coil is wrapped around the laser tube in order to confine the plasma into a smaller space and therefore increasing the argon ion density, but it is not necessary for laser output.

The $Ar^+$ laser can lase at many lines at once based on the cavity optics and can produce power outputs continuously at tens of watts. Some of the most common output wavelengths are blue at 457 nm, blueish-green at 488 nm, ultraviolet at 315, 514.5 nm, and many others including the infrared. The laser produces a significant amount of heat within the ceramic tube and therefore needs external cooling systems of either flowing air or water.

### 5.7.2 The Krypton Ion Laser

The krypton (Kr) ion laser works very much like the $Ar^+$ laser but with ionized krypton gas instead. This laser has multiple output wavelengths like the $Ar^+$ laser. In fact, there are so many lines available with the $Kr^+$ laser that the optics within the cavity can be designed in such a way as to generate an apparent "white-light" output by combining many of them. This laser is most commonly used for laser light shows.

## 5.8 More Exotic Lasers

### 5.8.1 The Free-Electron Laser

A free-electron laser or FEL is basically an electron accelerator in which the path of the electrons passes through an alternating magnetic field as shown

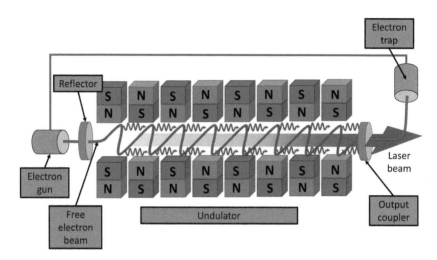

**FIGURE 5.31**
The FEL uses an electron accelerator to generate a laser beam.

in Figure 5.31. The Lorentz force of the charged particles passing through the magnetic field causes the electron's path to be curved. As the electron turns the curve a photon is emitted. The magnetic field generation system is called an "undulator" and it consists of dipole magnets alternating field polarity periodically as shown in the Figure 5.31.

The electrons are fired from an electron gun and accelerated to almost the speed of light. Once the beam passes into the undulator section the path of the electrons oscillate about the axis of propagation generating *synchrotron radiation*, which are incoherent photons. Using an optical cavity and/or a seed laser as an input source the field inside the cavity will force the electrons into bunches that in turn cause them to act as a density or current of electrons that begin to emit coherent photons at the same wavelengths.

The FEL is highly scalable and has been demonstrated to operate at wavelengths from the 1 mm range all the way out to X-rays in the fractions of nanometers. The wavelength is tuned by the magnetic field periodicity in the undulator, the electron energy density, seed laser properties, and the cavity optics. Research is being conducted to scale the output power of FELs up into the multi-megawatt class.

### 5.8.2 The Nuclear Bomb-Pumped Laser

Figure 5.32 shows the concept for pumping a disposable laser rod with soft x-rays from a nuclear explosion. A rod made of something like zinc would be exposed to the intense irradiance of soft x-rays and would become ionized and form a plasma of excited ions at an extreme temperature. Before the rod becomes vaporized, it is then super-radiant and emits laser light in the x-ray region for a very brief pulse.

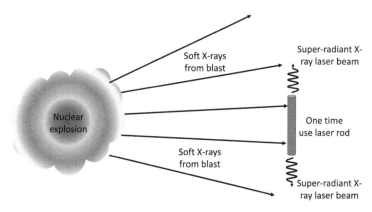

**FIGURE 5.32**
The nuclear bomb-pumped x-ray laser could produce enormously intense x-ray laser beams.

The pulse would be extremely energetic with estimations of as much as $20\,kJ/cm^2$ or more is possible. An optical cavity would be difficult as there is no technology currently available to act as mirrors for the x-ray beam. Also, note that x-rays do not propagate through the atmosphere well and this would be a laser system that needs to operate in a vacuum (in other words, space).

### 5.8.3 The Positronium Gamma Ray Laser

Positronium (Ps) is an element made of an antielectron (positron) orbiting an electron before they annihilate each other and generate a gamma ray. Positronium exists for less than a millionth of a second but could possibly be created in mass quantities by trapping positrons in large amounts and then having those positrons released onto a tube filled with many electrons at once (see Figure 5.33). The positrons would quickly create Ps with electrons

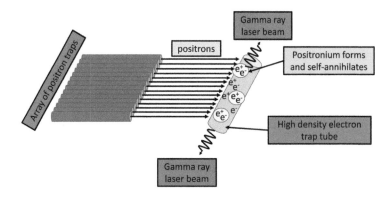

**FIGURE 5.33**
The positronium gamma ray laser could produce enormously intense gamma ray laser beams through matter–antimatter annihilation.

in the tube creating a population inversion. All of the Ps atoms would quickly annihilate themselves within the same microsecond or so releasing a very large gamma ray pulse.

The Ps gamma ray laser would actually be light amplification by the "annihilation" of matter and antimatter pairs, but it is still called a "laser." Also, the laser is in very basic research stages currently but in the not too distant future it is possible that the Ps gamma ray laser could be scaled up to a very high power repetitively pulsed laser system.

## 5.9 Chapter Summary

In this chapter, we learned that there are many types of lasers. In fact, there are so many different types of lasers we could not realistically discuss them all here. However, we did go through most of the general types of lasers and then discussed various specifics about them by giving examples of specific lasers.

We started in Section 5.1 with solid-state lasers where we discussed details of flashlamp-pumped and laser-pumped solid host materials. We learned about *second-harmonic generation* and how an infrared laser can have a green output beam. Then in Section 5.2 we studied the dye laser which is one of the most versatile, tunable, and robust lasers available. We learned that the dye can be dissolved in a solvent like methanol or it can be doped into a solid host like plastic.

In Section 5.3, we discussed the semiconductor laser briefly where we learned the difference between a standard diode laser that emits light orthogonal to the electrical current path and a VCSEL which lases along the same axis as the current flow.

In Section 5.4, we discussed gas lasers. There are many types of gas lasers operating as pulsed and/or as continuous wave lasers. We learned about one of the "work horse" lasers of the community—the HeNe. As the HeNe is used in almost every laser laboratory and by almost every laser scientist and engineer it was good to become aware of how this device operates and what it's capabilities are. We also discussed the many other types of gas lasers including the gas dynamic laser that is very similar to the chemical lasers discussed in Section 5.5.

We finished up by discussing metal vapor lasers in Section 5.6, ion lasers in Section 5.7, and then some more exotic ideas in Section 5.8. Again, the lasers discussed in this chapter are only a few of the lasers that exist and new ones are being invented all the time. While we did not exhaust every possibility of laser configurations here, it was a very good overview and general description of most of the better known and widely used lasers.

## 5.10 Questions and Problems

1. The ruby rod contains what which provides the stimulated emission transitions for laser action?
2. What is the difference in a "linear" and "helical" flashlamp?
3. What is *Q-switching*?
4. What type of optical device must be used for *Q-switching*?
5. What is "yelf"?
6. What is YAG?
7. What process is used to convert an infrared beam to a green beam?
8. What is KTP?
9. What is the more precise chemical formula for KTP?
10. What is BBO?
11. What are fiber laser-active medium fibers typically doped with?
12. What is Rhodamine 6G?
13. What is a *coaxial* flashlamp?
14. Why are dye lasers tunable across many wavelengths?
15. Why are dye lasers typically pulsed?
16. Why is it so difficult to have continuous wave dye lasers?
17. What is meant by "forward bias"?
18. What is a p-type material?
19. What is the ratio of helium to neon in a HeNe laser?
20. What is a TEA laser?
21. Give two examples of TEA lasers.
22. What is an excimer laser?
23. What is an exciplex laser?
24. Why are deuterium fluoride (DF) lasers typically of more interest than HF lasers for terrestrial applications?
25. Why do metal vapor lasers typically self-quench?
26. How can a krypton ion laser be "white light"?
27. What is an "undulator"?
28. Why do x-ray lasers typically not have mirrors?
29. How are FELs tunable?
30. What is positronium?
31. Explain lasing by annihilation?

# 6

## *How Do We Describe Lasers?*

From the beginning of this book, we have been learning the science behind what lasers are. Now that we have a fairly good understanding of how a laser operates from a science point of view, it is time to be more pragmatic. Knowing how to draw a laser on paper and write down rate equations is one thing, but to actually make use of lasers a good scientist or engineer also needs to know a little more practical knowledge about them.

In this chapter, we will learn details about what makes a specific laser what it is. We will learn how to describe and characterize lasers through measurements that can be made in the laboratory that will tell us what the laser might be used for. For example, if there is a need for a "narrow linewidth green laser" just how narrow is narrow and how green is green? How do we know the linewidth of a laser and how do we know specifically what color it is? We make measurements using specific devices.

As we have seen in Chapter 5, lasers come in all shapes, sizes, and configurations. The key for researchers and engineers is to know enough about specific lasers to meet specific tasks. Some tasks require pulsed beams and some require continuous ones. Some require high power and some single photons. Some experiments or applications might require specific pulse shapes or linewidths or wavelengths. This chapter will give us a better understanding of how to describe lasers with enough detail to choose them for specific applications.

## 6.1 Continuous Wave or Pulsed

### 6.1.1 Continuous Wave Laser Beams

When it comes to lasers that produce continuous output beams, we call them "continuous wave" and more commonly "CW." CW lasers, like the HeNe for example, have a steady stream of output light once laser action is initiated that will continue until the power is turned off. In fact, the HeNe was the first CW laser ever demonstrated.

Figure 6.1 shows three laser pointers with CW beams. Since the beams are always on, the beam can always be seen if there is some media to scatter some of the laser beam light such as smoke, dust, or fog. If we looked at the output power of the lasers with a power meter the measurement would read a continuous nonzero value.

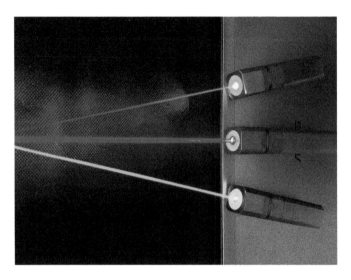

**FIGURE 6.1**
CW laser pointers have beams that are always on. (Courtesy Pangkakit at Chinese Wikipedia GNU Free Documentation License.)

## 6.1.2 Pulsed Laser Beams

Lasers that output light discontinuously are called "pulsed lasers." Pulsed lasers of all *pulse widths* can be found. The laser pulses can be generated by multiple means. Some lasers simply track the input power pump light pulse. This is known as *gain switching*. Some lasers are *Q-switched* as discussed in previous chapters. Some lasers use a technique called *modelocking*, which uses active optical elements within the laser cavity to limit the time, and sometimes wavelength, that a beam can be propagating within it.

Figure 6.2 shows a typical Nd:YAG *Q-switched* laser pulse. The pulse is typically measured by a fast photomultiplier tube or photodetector that is connected to a fast oscilloscope. The *pulse width* is measured at full-width half-maximum (FWHM) value. The pulse is about 7.5 ns wide.

Figure 6.3 shows three methods for measuring the laser pulse's energy (typically arbitrary or relative scale) as a function of time in the laboratory. This measurement is the *temporal profile* of the laser pulse. The most obvious technique is shown in Figure 6.3a and is the *direct measurement* technique. This technique is used to directly shine the laser beam into the photodetector. However, many photodetectors are easily damaged at higher powers so this technique is only recommended for lower-power laser pulses. Figure 6.3b shows the *specular measurement* technique which is used to measure a specular reflection off the first surface of a beam splitter which is sometimes for very high-power lasers called a *scraper* instead of a beam splitter and usually only reflects single-digit percentages and transmits the rest. A common technique for medium power visible lasers is to use a glass microscope slide

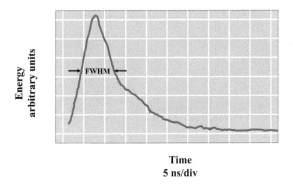

**FIGURE 6.2**
A typical output pulse shape for a Nd:YAG Q-switched laser. The pulse width is about 7.5 ns at FWHM value.

as a *scraper* mirror, as it is only 4% reflective on the first surface and transmits the rest. *Scraper* mirrors are also sometimes used for output couplers in cavities with very highly reflective mirrors on each end. Also, sometimes a *scraper* is 100% reflective (or as near as possible) at the lasing wavelength and is placed partially into the beam to "scrape" off some of the photons for measurement or use.

Figure 6.3c shows a more indirect technique called the *diffuse measurement* technique. In this case, the beam is incident on a diffuser screen, sometimes just a piece of paper, cardboard, or even the wall, where it is diffusely reflected. The photodetector is aimed at the spot on the screen and captures diffusely reflected photons.

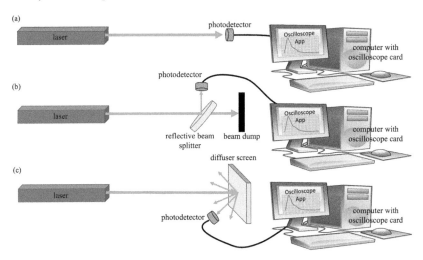

**FIGURE 6.3**
Methods for detecting laser pulses and measuring their temporal profile in the laboratory. (a) Direct measurement, (b) specular measurement, and (c) diffuse measurement.

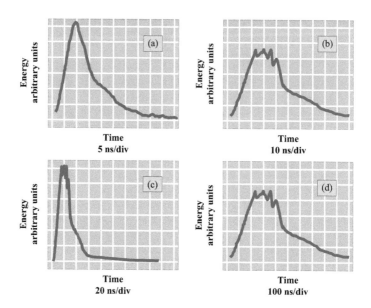

**FIGURE 6.4**
For Example 6.1 measure the pulse width of the above laser pulses.

**Example 6.1: Measuring the *Pulse width* of a Laser**

Consider the waveforms shown in Figure 6.4a–d. Measure the *pulse width* at FWHM. The peak of Figure 6.4(a) is about six divisions plus a little bit high and therefore the half-maximum is at about three divisions (plus a bit). The FWHM *pulse width* is about 10 ns. The other values are (b) the *pulse width* at FWHM is about 35 ns, (c) ~20 ns, and (d) 350 ns.

Note that measuring with a ruler application or actually using a real ruler will allow for slightly more accurate measurements. However, there are always errors in such measurements as the thickness of the trace causes uncertainty, the accuracy of the ruler marks will cause some uncertainty, and plain old human error will induce some. That being said, these are all error types that can be accounted for and should only be a few percent or so uncertainty. Typically, that will be precise enough for the laser experiment or application.

## 6.2 Laser Modes

### 6.2.1 Longitudinal Modes

Laser cavities, as we have learned, are resonators. A typical two-flat mirror laser cavity resonates at specific frequencies based on the separation distance, $L$, between them. These resonant frequencies are standing waves oscillating between the mirrors, like a guitar string oscillates when plucked. Like the guitar string, the wave amplitude must be zero at the boundaries and the number of resonant frequencies that can oscillate within the cavity, $q$, known as the *mode number* is

$$q = \frac{2L}{\lambda} = \frac{2Lv}{c} \tag{6.1}$$

where $\lambda$ is the wavelength of the laser and $v$ is the frequency. We should note here that we can rearrange Equation 6.1 and realize that $q$ must be an integer value where

$$L = q\frac{\lambda}{2}. \tag{6.2}$$

Equation 6.2 tells us that the length of the resonant cavity must be integer multiples of the wavelengths of the resonant cavity light divided by two. These resonances are called *longitudinal modes*. Figure 6.5 shows the first six of these modes that are allowed within a cavity of arbitrary length. Figure 6.6 shows a log-log graph of the number of modes versus the cavity length as given by Equation 6.1. Notice that there are millions of modes that can exist as the cavity length increases because the wavelength of the laser light is very small in comparison.

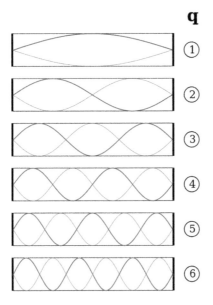

**FIGURE 6.5**
First six longitudinal modes of a resonant cavity.

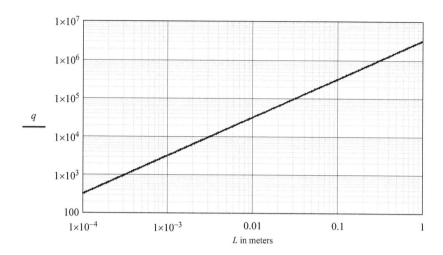

**FIGURE 6.6**
The number of longitudinal modes as a function of cavity length.

While a cavity can have many modes, they must be spaced apart from each other in frequency as

$$\Delta \upsilon = \frac{c}{2nL} \qquad (6.3)$$

or in wavelength as

$$\Delta \lambda = \frac{c}{\nu_o + \dfrac{c}{2nL}} - \lambda_o \qquad (6.4)$$

where $\nu_o$ is the center lasing frequency, $\lambda_o$ is the center lasing wavelength, and $n$ is the index of refraction of the laser medium (we will simplify for now and assume it $n \sim 1$ which is not always the case as laser rods are often $n > 1$). Figure 6.7 shows a graph of Equation 6.3 as a function of cavity length. As the cavity length increases, the separation between allowed modes decreases.

Consider a HeNe laser with a cavity length of 30 cm. The HeNe gain curve is about 1,400 MHz wide at FWHM as shown in Figure 6.8. According to the graph in Figure 6.7, a 30-cm-length cavity has a longitudinal mode separation of about 500 MHz. Therefore, only four modes at best can exist in the laser. For shorter laser cavity length, it is likely only two longitudinal modes can exist. Using Equation 6.4 for the HeNe and assuming the laser central wavelength is 632.8 nm ($4.74 \times 10^{14}$ Hz) then the wavelength spacing for the longitudinal mode is about 0.0005 nm.

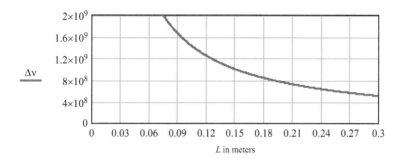

**FIGURE 6.7**
The longitudinal mode separation, $\Delta v$, as a function of cavity length, $L$.

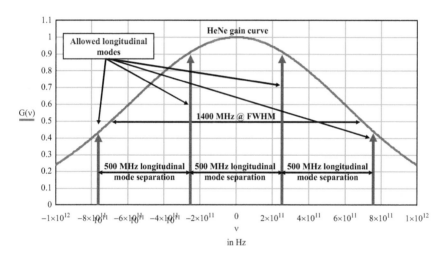

**FIGURE 6.8**
The allowed longitudinal modes of a HeNe laser with a 30-cm cavity length.

### 6.2.1.1 Longitudinal Modes in Pulsed Lasers

Using the technique shown in Figure 6.3 the *temporal profile* of a coaxial flashlamp-pumped liquid dye laser using the dye Rhodamine 6G dissolved in methanol as the active medium. The resonant cavity shown in Figure 6.9 was more complex than standard flat mirrors. There was a flat output coupler mirror and on the back-reflector side, there was an arrangement of prisms and a grating known as a "multi-prism Littrow grazing incidence grating" configuration. The prisms and gratings were used to spread out the broadband laser light and to limit the band which could lase and in turn generate a narrow linewidth output laser beam that was tunable from about 565–610 nm by adjusting the incidence angle of the grating. The overall cavity length, $L$, was about 50 cm.

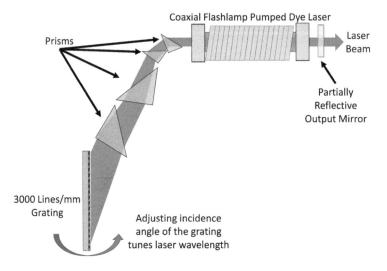

**FIGURE 6.9**
The multi-prism Littrow grazing incidence grating cavity for a coaxial flashlamp-pumped dye laser is used to create a narrow linewidth laser beam.

### Example 6.2: Measuring the Longitudinal Modes

Figure 6.10 is the temporal profile measurement of the dye laser in the multi-prism arrangement as discussed above. The vertical axis is relative energy output of the beam (or irradiance since the value is relative and not exact) and the horizontal axis is time in 20 ns per big division. There is a sinusoidal signal that is clearly following the output pulse.

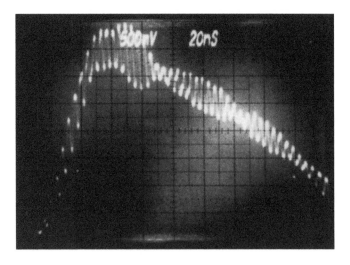

**FIGURE 6.10**
The multi-prism Littrow grazing incidence grating coaxial flashlamp-pumped dye laser temporal profile.

The sinusoidal aspect of the waveform is also growing and shrinking in amplitude following a second sinusoidal envelope. The period of this sinusoidal envelope is the *beat frequency* between two longitudinal modes. The envelope peak-to-peak time measured from the picture is about 30 ns, which is the period of the *beat frequency*. The *beat frequency* corresponds to the mode separation and is calculated as the inverse of the period to be about 333 MHz. This tells us that there were two longitudinal modes operating within the laser cavity separated by about 333 MHz from each other or about 0.0004 nm.

## 6.2.2 Transverse Modes

In the previous section, we discussed how the laser output energy was distributed over time and frequency within the cavity. We now understand what the temporal mode of a laser is and how to measure it. But what about the beam's cross-sectional profile in space? In other words, how do we measure and describe the *spatial profile* of the laser beam?

Figure 6.3 shows three methods for measuring the laser beam's energy (typically arbitrary or relative scale) as a function of spatial dimension in the laboratory. This measurement is the *spatial profile* of the laser beam. The most obvious technique is shown in Figure 6.11a and is the *direct measurement* technique. This technique is used to directly shine the laser beam into the camera and to capture the image of the beam. However, many cameras are easily damaged at higher powers so this technique is only recommended for lower-power laser beams. Note that the application of attenuation filters can also be implemented to reduce the irradiance of the laser beam to safe levels for the camera equipment.

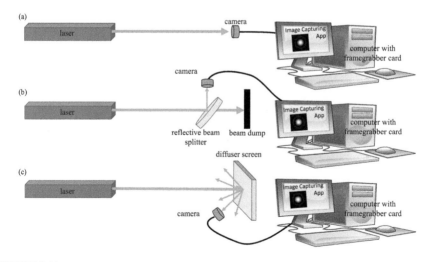

**FIGURE 6.11**
Methods for detecting laser pulses and measuring their spatial profile in the laboratory. (a) Direct measurement, (b) specular measurement, and (c) diffuse measurement.

Figure 6.11b shows the *specular measurement* technique which is used to measure a specular reflection off the first surface of a beam splitter. Most first-surface reflections are typically single-digit percentages of the entire beam's output and will often be enough of a reduction to be safe for the camera. Also, again we can implement attenuation filters if needed.

Figure 6.11c shows a more indirect technique called the *diffuse measurement* technique. In this case the beam is incident on a diffuser screen, sometimes just a piece of paper, cardboard, or even the wall, where it is diffusely reflected. The camera is focused on the screen and captures the image of the laser beam spot. This technique is typically the safest in regards to the camera equipment; however, in actuality, it induces an angular component to the image of the laser beam as it is at an angle to the screen. In some cases, the laser beam can be incident on a very thin screen and the camera can be in direct line with the beam on the opposite side of the screen in order to remove the angular issues with measurement.

### 6.2.2.1 Transverse Electromagnetic Modes

The laser scientist or engineer will often see in the literature (or hear in the lab) the description of a laser beam's spatial cross section in terms of *transverse electromagnetic modes* or TEMs. This terminology comes from classical electromagnetics theory describing electromagnetic (EM) fields in waveguides and for laser cavities. The number of spatial modes in the beam profile are usually denoted by subscripts for $x$ and $y$ for rectangular cavities and $r$ and $\phi$ for cylindrical cavities. The mode numbers represent how many separations in the beam cross section are there.

Figures 6.12 and 6.13 show rectangular and cylindrical cavity $\text{TEM}_{xy}$ and $\text{TEM}_{r\phi}$ modes, respectively, for the first nine modes. Most laser scientists and

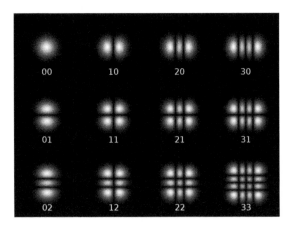

**FIGURE 6.12**
The first nine TEMs for a rectangular cavity. (Courtesy Dr. Bob Wikimedia GNU Free Documentation License).

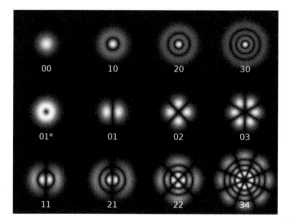

**FIGURE 6.13**
The first nine TEMs for a cylindrical cavity. (Courtesy Dr. Bob Wikimedia GNU Free Documentation License).

engineers strive to achieve laser output profiles with $TEM_{00}$ modes when wanting optimal energy per area in a beam. However, some applications might require having beam shapes with different profiles and the laser cavity might be designed to generate a specific spatial pattern.

## 6.3 Spectral Content

### 6.3.1 Linewidth

We have discussed laser *linewidth* throughout this book and so we are familiar with the concept. The spectral content of the laser beam is determined from knowing the *coherence length* of the laser cavity, $L_c$, which is

$$L_c = \frac{c}{n\Delta v} = \frac{\lambda_o^2}{n\Delta\lambda}. \tag{6.5}$$

The *coherence length* of the cavity is the propagation distance at which the light beam maintains the property of beam or light field self-interference. In other words, at distances within the *coherence length* the light inside the cavity will interfere and diffract strongly. The *linewidth* can be found as a function of the *coherence length* by

$$\Delta v \approx \frac{c}{L_c} \tag{6.6}$$

and

$$\Delta\lambda \approx \frac{\lambda_o^2}{L_c}. \tag{6.7}$$

**Example 6.3: Calculate the *Linewidth* of a Laser**

Consider a laser with a coherence length of about 90 cm and a lasing wavelength of 590 nm. Use Equation 6.7 to determine the *linewidth*.

$$\Delta\lambda \approx \frac{\lambda_o^2}{L_c} = \frac{\left(590\times 10^{-9}\,\text{m}\right)^2}{0.9\,\text{m}} = 3.8677\times 10^{-13} \approx 0.0004\,\text{nm}. \tag{6.8}$$

### 6.3.1.1 Measuring the Linewidth of a Laser

Calculating the linewidth of a laser is one thing, but how is it measured in the laboratory?

The linewidth can be measured in a couple of ways: (1) a device called a *monochromator* can be used and (2) a device called a Fabry–Perot *etalon* can be used. A *monochromator* implements optically dispersive components such as prisms and gratings to spread a beam out into its multiwavelength components much in the same way a prism separates the colors of sunlight into the rainbow spectrum. Figure 6.14 shows how a *monochromator* works. In order to measure the linewidth of a laser beam the prism is rotated in one direction until no signal is detected and then measurements are made as the prism is rotated through the peak of the signal and then to the other side where it is

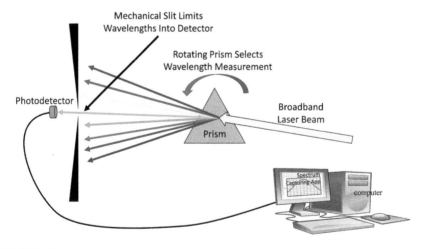

**FIGURE 6.14**
The monochromator is used to measure specific wavelengths of light and can be used to measure linewidth of a laser beam.

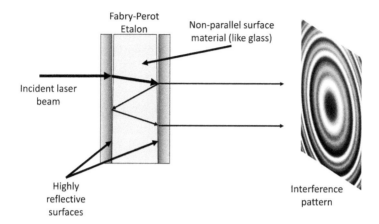

**FIGURE 6.15**
An etalon is used to measure linewidth of a laser beam.

zero. The *monochromator* must be calibrated with a known spectral source in order for the measurements to be in real (not relative) units.

Figure 6.15 shows a Fabry–Perot *etalon*. *Etalon* is the French word for "measuring gauge" or "standard." The device was developed by Charles Fabry and Alfred Perot in 1899. It consists of a wedge-shaped or non-parallel surface optically transmissive material (like glass) with highly reflective coatings or mirrors in intimate contact with each surface. The mirrors are designed to be parallel with each other. As light passes into one side of the *etalon* some of it is reflected back and forth with each pass. The exiting light, therefore, is slightly phase delayed with each other's pass and the different phase-delayed light beams interfere with each other generating a pattern of concentric rings as shown in Figure 6.16.

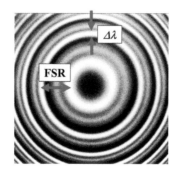

**FIGURE 6.16**
The FSR and the linewidth as measured with an etalon. (Image of rings (modified) Courtesy of John Walton Wikipedia Creative Commons Attribution 3.0 Unported license.)

The distance between two adjacent rings is known as the *free spectral range* (FSR) of the etalon. The width of each ring is the *linewidth* of the light entering the etalon. The linewidth of the input light is calculated as

$$\Delta\lambda = \frac{\text{FSR}}{\mathcal{F}} \tag{6.9}$$

where $\mathcal{F}$ is the finesse of the *etalon* and is approximated as

$$\mathcal{F} \approx \frac{\pi R^{\frac{1}{2}}}{1 - R^2} \tag{6.10}$$

and $R$ is the reflectance of each mirror (assuming they are equal). Note that Equation 6.9 will often be rewritten as frequency units

$$\Delta\upsilon = \frac{\text{FSR}}{\mathcal{F}}. \tag{6.11}$$

The FSR will be given in either units of length or frequency (such as nm or MHz).

In order to use the *etalon* to make measurements of *linewidth* we must know information about the specific device being used in order to calibrate it. Typically, the manufacturer or vendor will sell an etalon with a specified calibrated FSR and *finesse*. The laser scientist or engineer then must use those numbers in comparison with measured numbers to make actual physically calibrated *linewidth* measurements.

### Example 6.4: Using an *Etalon*

Assume the image shown in Figure 6.16 is measured with a ruler and that $\text{FSR}_{\text{measured}} = 2\,\text{cm}$ and that $\Delta\lambda = 0.5\,\text{cm}$. The manufacturer's specifications for the *etalon* used is that $\text{FSR}_{\text{spec}} = 1{,}500\,\text{MHz}$ and the finesse is 4. Determine the actual linewidth of the beam.

$$\Delta\upsilon_{\text{measured}} = \frac{\text{FSR}_{\text{measured}}}{\mathcal{F}}. \tag{6.12}$$

$$\Delta\upsilon_{\text{calibrated}} = \frac{\text{FSR}_{\text{spec}}}{\mathcal{F}}. \tag{6.13}$$

Divide Equation 6.13 by 6.12 to get

$$\frac{\Delta\upsilon_{\text{calibrated}}}{\Delta\upsilon_{\text{measured}}} = \frac{\dfrac{\text{FSR}_{\text{spec}}}{\mathcal{F}}}{\dfrac{\text{FSR}_{\text{measured}}}{\mathcal{F}}}. \tag{6.14}$$

Rearranging Equation 6.14

$$\Delta\upsilon_{\text{calibrated}} = \Delta\upsilon_{\text{measured}} \frac{\text{FSR}_{\text{spec}}}{\text{FSR}_{\text{measured}}}. \tag{6.15}$$

Inputting the numbers into Equation 6.15 results in

$$\Delta v_{\text{calibrated}} = 0.5\,\text{cm}\,\frac{1,500\,\text{MHz}}{2\,\text{cm}} = 375\,\text{MHz}. \qquad (6.16)$$

Note that since the measured FSR and *linewidth* were in the same units (cm) $\Delta v$ is interchangeable with $\Delta \lambda$ here but only for the measured values.

We should point out here that the Fabry–Perot *etalon* shown in Figure 6.15 can also be constructed with an open-air gap between the reflective surfaces rather than a glass wedge. This configuration is typically called a Fabry–Perot *interferometer*. We should also realize that this interferometer is just an optical cavity with two highly reflective flat mirrors on each end, which is pretty much the description of a basic laser cavity. We now have the background to realize that most flat mirror lasers are indeed Fabry–Perot cavities.

## 6.3.2 Tunability

As shown in Figure 6.9 and briefly discussed throughout this book, many lasers are tunable across many wavelengths. The dye laser has a very broadband gain curve and, in fact, takes some significant effort to generate narrow linewidth pulses. The HeNe laser is inherently narrow linewidth, but we did mention that it could lase at several different laser wavelengths. We have mentioned transition selection but only suggested this is done with specifically coated or designed cavity optics.

Most laser transitions that can be selected by tuning the gain curve is done so by limiting the gain where lasing is not desired and by maximizing it where it is desired. Some lasers can be tuned across very broad portions of the spectrum while others are limited to narrow bands. One of the most tunable is the dye laser. Figure 6.17 shows a tunable cavity configuration for a coaxial flashlamp-pumped dye laser. A standard flat broadband partially reflective output coupler is used, but the back-reflector side of the cavity consists of a

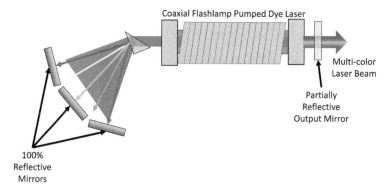

**FIGURE 6.17**
A laser cavity configuration that generates multiple laser beams at multiple colors.

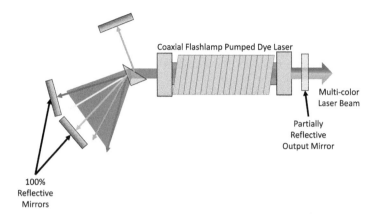

**FIGURE 6.18**
A laser cavity configuration using first-surface reflection and dispersive elements to create multiple lasing lines.

dispersive element like a prism that spreads the spectral line pathways out in spatial dimension much in the same way the monochromator does (see Figure 6.14). As the pathways are separated by the dispersive element, a 100% reflector for each desired wavelength is placed in the appropriate location. In this way, multiple wavelengths from one laser can be achieved.

In reality, it is difficult to achieve more than two to three lasing lines due to physical dimensional constraints and due to there only being so much gain within the cavity for each spectral mode. Figure 6.18 shows a second configuration for tuning and multiwavelength operation of a dye laser by using a first-surface reflection path off the prism. The optic can be coated to maximize the desired wavelength's reflection and to increase the cavity gain for that particular pathway. Again, however, there is only so much gain in the cavity available for the multiple lines.

Figure 6.19 shows a configuration for single line tuning. The incidence angle for a single reflector can be changed to tune the laser across the

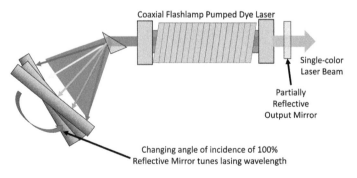

**FIGURE 6.19**
The typical tunable laser cavity configuration.

available gain spectrum. A second alternative would be to rotate the prism. There are many applications for tunability such as tuning the laser source across an absorption spectrum of a molecule to determine the identity of the molecule based on fluorescence and absorption.

## 6.4 Collimation, Divergence, Beam Expansion, and Beam Reduction

Another characteristic of all lasers is the *divergence* of the output beam. As a laser beam leaves the output coupler it will spread as it propagates. In other words, the spot diameter will get larger as the beam travels farther from the laser as shown in Figure 6.20. The *full-angle beam divergence* is calculated by

$$\Theta = 2\arctan\left(\frac{D_f - D_i}{2\Delta d}\right). \tag{6.17}$$

where $D_f$ is the final spot diameter, $D_i$ is the initial spot diameter, and $\Delta d$ is the distance between the final and initial spots. Typical divergences for HeNe lasers, for example, are usually measured in the milliradians.

All laser beams diverge naturally as a consequence of propagation and diffraction. Optical elements can be used to focus beams but there are limits to the control of beam divergence simply due to the physics of propagation. A common practice for laser scientists and engineers is to "collimate" a laser beam using a telescope or other dispersive optical element system. It is often misunderstood by the novice laser operator that a "collimated" laser beam will not remain collimated forever because of diffraction.

Figure 6.21 shows a typical configuration for expanding and collimating a laser beam using two lenses and a spatial filter (a pinhole). The two lenses are placed apart from each other such that their focal spots coincide and in the near field within a meter or two from the lenses the beam appears to be perfectly collimated with no divergence. While this may be true (no divergence) diffraction due to propagation is still happening and in the far-field

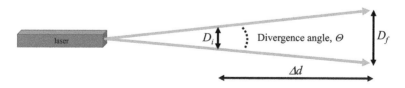

**FIGURE 6.20**
The divergence of the laser beam can be calculated based on the spot diameter at two separate distances.

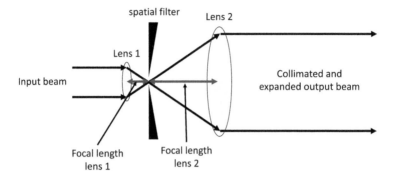

**FIGURE 6.21**
The typical laser beam collimation and beam-expanding optical system configuration. Note: If input beam is not collimated lenses will not be focal lengths apart.

distances (greater than a thousand meters or so) the beam size is limited by the output lens aperture diameter, $D_{aperture}$, and the distance propagated, $z$, as

$$D_{spot} = \frac{2.44 \lambda z}{D_{aperture}}.$$ (6.18)

Figure 6.21 also shows *beam expansion*, which is the increase in beam diameter from input to output of the optical system. Many laser beams exit the laser aperture in a very tiny beam only a few millimeters in diameter and for many applications this beam needs to be expanded many times. The expansion of the beam expander shown in Figure 6.21 is calculated as

$$MP = \frac{f_2}{f_1}$$ (6.19)

where $f_1$ and $f_2$ are the focal lengths of lenses 1 and 2, respectively, and MP is the *magnification power* of the beam expander. We also should realize that the *beam expander* can be used in reverse as a *beam reducer*.

## 6.5 Output Energy, Power, Irradiance, and Beam Quality

### 6.5.1 Instantaneous versus Average

One main characteristic of any laser that every laser scientist and engineer must know and truly understand is the amount of output energy, power, and irradiance the device supplies. Every laser application requires a certain type of output and in order to implement safety protocols properly the output must be truly understood.

CW lasers are typically described by their *average power* output measured in watts. The beam itself being emitted is typically described by irradiance in watts per meter squared (often times irradiance is given as watts per centimeter squared and we discussed in Chapter 1 how some text, papers, scientists, and engineers confusingly call it *intensity*). CW laser output power can range from microwatts (maybe lower) up to megawatts and maybe even higher in the future.

Pulsed lasers are often described by their output energy per pulse. It is very common to see Nd:YAG Q-switched laser pulses in the many joules per pulse range and they can reach upwards into the kilojoule range. Many flashlamp-pumped dye lasers often produce millijoule to joule per pulse output energies. The power output of a pulsed laser must be described in two ways: (1) instantaneous power or power per pulse and (2) average power or power over many pulses.

Figure 6.22 shows a chain of pulses from a laser as measured by an energy meter. There are four pulses in the time frame displayed. Each pulse is pretty much the same as other pulse with a peak height at about 14 mJ and a *pulse width* of about 7 ns at FWHM. The pulses are being repeated periodically with some *pulse repetition frequency* (PRF) with a period of $\tau_{PRF}$ of about 30 ns. The PRF is found as the inverse of the period

$$PRF = \frac{1}{\tau_{PRF}}. \tag{6.20}$$

So, the PRF for the pulses in Figure 6.22 is about 33.3 MHz.

The energy per pulse is the area under one pulse, which is about 140 mJ or so. The area under the graph can be computed with a computer program or

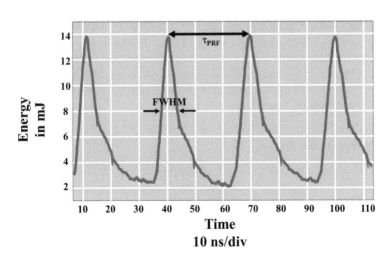

**FIGURE 6.22**
A chain of laser pulses with PRF of 33.3 MHz.

graphically. In most cases, an energy meter will do the integration for us. The *instantaneous power* is the power per pulse which is found as

$$P_{\text{inst}} = \frac{\mathcal{E}_{\text{pulse}}}{\tau_{\text{PRF}}}.$$ (6.21)

For the pulses in Figure 6.22 the *instantaneous power* is about 4.6 MW. That sounds amazingly high, but recall that it is only a 140 mJ total energy in the pulse. The pulse is just very very short.

It is the *average power* that is more of a comparison to a CW laser power. The *average power* tells us the amount of power being supplied by the laser pulse train over time. Referring back to Figure 6.22 and assuming that this train of pulses will continue as long as power is supplied to the laser then we can determine the *average power* output, $P_{\text{av}}$, to be

$$P_{\text{av}} = \frac{\tau_{\text{pulse}}}{\tau_{\text{PRF}}} P_{\text{inst}} = \tau_{\text{pulse}}\text{PRF}\,P_{\text{inst}}.$$ (6.22)

We should also define here the *duty cycle* which is the percentage of time that the laser is emitting power during operation. The duty cycle, $\delta$, is

$$\delta = \frac{\tau_{\text{pulse}}}{\tau_{\text{PRF}}} = \tau_{\text{pulse}}\text{PRF}.$$ (6.23)

So, Equation 6.22 becomes

$$P_{\text{av}} = \delta P_{\text{inst}}.$$ (6.24)

Figure 6.23 shows a graph of the *average power* as a function of PRF for the pulses given in Figure 6.22. With the PRF at 33.3 MHz we see that there is still a significant average power output, but if we changed the PRF to a more realistic 1 Hz to 1 kHz range the *average power* would be between 100 mW to 100 W. Many Nd:YAG lasers can provide this type of energy per pulse at PRFs ranging from hertz to kilohertz.

### 6.5.2 Power in the Bucket

For many lasers, especially very high-energy ones, the beam tends to take on unusual spatial profiles with propagation. It is sometimes unclear how to define the energy or power within the beam area if that area is of an odd shape. A typical measurement made in this type of case is the so-called "power in the bucket" or PIB measurement.

The PIB measurement is defined by the laser scientist or engineer as a specific-sized "bucket" or aperture that is typically circular. Once the

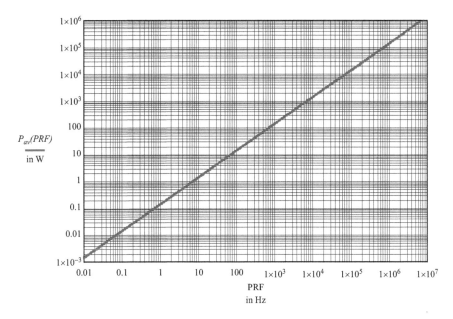

**FIGURE 6.23**
The average power of pulses from Figure 6.22 as a function of PRF.

"bucket" size is defined then only energy that is encircled within the bucket is accepted for measurement purposes. In fact, a physical "bucket" is often just the aperture stop or entrance aperture to the power meter. In some cases, an actual bucket aperture is implemented and based on the power levels being measured can range from a piece of cardboard with a hole cut in it to very thick steel plates with holes.

The measurement actually requires a calibration at or near the exit aperture of the laser where the same "bucket" is used to measure the beam's power. Then the "bucket" is used to take power measurements a predefined distance from the laser and the two measurements are compared in a ratio. A PIB ratio of 1 would be most ideal, meaning that all the power leaving the aperture of the laser made through the propagation path are still intact and into the same-sized "bucket" down range. The PIB measurement is useful in describing a specific laser's ability to beam power over some distance to a receiver or target and is important for beaming power, optical communications, and laser weapon applications.

### 6.5.3 Beam Quality

Another characteristic of lasers that the scientist or engineer will often hear is the so-called "beam quality" of the laser beam itself. There are many definitions for beam quality and, in fact, beam quality can even be application specific. Anthony E. Siegman first proposed that there should be a *beam quality factor*, $M^2$, which describes the smallest spot size for a beam as well as the beam's divergence.

Siegman's *beam quality factor* assumes that the best case for a laser beam is to be a diffraction-limited Gaussian beam. This means that the beam's spatial profile exiting the laser is a smooth Gaussian function and that any propagation effects are simply due to classically described diffraction effects. $M^2$ is implemented within the formula for *half-angle beam divergence* of a Gaussian beam

$$\Theta_{\text{half}} = M^2 \frac{\lambda}{\pi w_o} \tag{6.25}$$

Where $\lambda$ is the wavelength of the light and $w_o$ is the radius at the smallest point of the Gaussian beam within the cavity and is called the *beam waist*. M is never smaller than unity and in essence describes how the beam will grow with propagation distance, which will impact the PIB measurement ratio. The larger $M$ will dramatically and negatively impact the PIB measurement.

Another beam quality parameter often discussed is the *beam parameter product* or BPP. The BPP is defined as

$$\text{BPP} = \Theta_{\text{half}} w_o. \tag{6.26}$$

Rewriting Equation 6.25 results in

$$\text{BPP} = M^2 \frac{\lambda}{\pi}. \tag{6.27}$$

It is difficult at first for the novice laser scientist or engineer to realize the meaning of these two beam quality concepts. After experience with actual laser systems, the idea of beam quality will begin to become more apparent. To better understand how beam quality as described here is useful consider a very high-power CW laser with standard flat mirrors. At low powers, say just above laser threshold, active gain medium is still relatively cool and looks uniform in index of refraction. If it is a solid-state laser, we could say the rod is uniform in index of refraction from one end to the other. At this point, the beam is likely to be very close to ideal beam quality and M will be very close to unity.

After several seconds of operation, much heat will have built up within the laser system including within the rod itself. It is very unlikely that the heat is perfectly uniformly distributed throughout the rod material and therefore where the temperature varies so will the index of refraction. Now, suddenly, our once perfect laser beam is passing through index of refraction variations which look like small and random-valued lenses. Our beam will then be spread out due to these fluctuations in index of refraction. The beam quality will suffer in return and therefore M becomes greater than one. As the laser power is increased, the thermal issues increase and the beam quality will continue to suffer.

## 6.6 Efficiency

### 6.6.1 Wall-Plug Efficiency

All lasers need a power source. They must get power from either a stored system such as batteries, an electrical outlet in the wall, solar panels, maybe a generator that runs off gas, or even from stored energy in nuclear or chemical systems. The point is lasers need power.

A very common parameter discussed by the laser scientist or engineer is the *efficiency* of the laser. There are many types of efficiencies to discuss when it comes to a laser and this topic is often confusing or at the very least obfuscated. In order to hopefully avoid confusion, we will start our discussion from where the power is first supplied—the wall-plug. Consider the laser system diagram shown in Figure 6.24. The *wall-plug efficiency*, $\eta_{\text{wallplug}}$, is the ratio

$$\eta_{\text{wallplug}} = \frac{P_{\text{laserout}}}{P_{\text{in}}}. \tag{6.28}$$

If a laser produces an average output power of $1\,\text{mW}$ and requires $100\,\text{mW}$ to operate then the *wall-plug efficiency* is 0.01% or 1%. Actual *wall-plug efficiencies* for lasers can range from single digit up to about 20% or so depending on the laser system. Some systems of diode-pumped solid-state lasers report upwards to 30% wall-plug efficiency but there are often questions of the calculations and if some things such as losses in the power supplies were left out. We will follow the philosophy herein that calculating or measuring

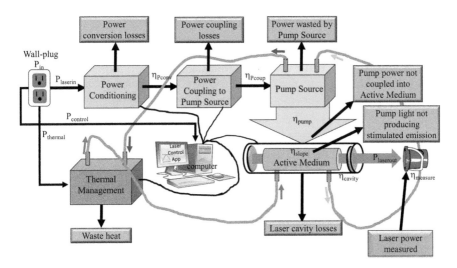

**FIGURE 6.24**
Power used by a laser system has many losses to be considered.

*wall-plug efficiency* requires accounting for every bit of power that is required from the wall-plug to operate the laser.

While it might seem difficult to understand all the losses of a system, it is possible to measure the voltage level and current draw (and therefore the power) needed to power all the systems connected to the laser as shown in Figure 6.24. We can start tracking the power, and therefore begin looking deeper into the losses in the system, by realizing that the laser system consists of three major subsystems: (1) the laser assembly subsystem, (2) the thermal management subsystem, and (3) the control subsystem. We can break Equation 6.28 down into these subsystems as

$$\eta_{\text{wallplug}} = \frac{P_{\text{laserout}}}{P_{\text{laserin}} + P_{\text{thermal}} + P_{\text{control}}}. \tag{6.29}$$

The thermal management subsystem is used to cool the laser-active medium, optics, and the pump source. Oftentimes, flashlamp-pumped lasers require a flowing liquid coolant (such as chilled water) and some other lasers require even more exotic thermal management. The power requirements for this subsystem typically involve connecting chillers, pumps, and temperature control circuits to wall-plug power. A power meter can be used to measure the actual usage, but calculations can be performed based on the thermodynamics of the specific laser system to reach a fairly accurate answer. Thermal analysis will be required for any laser scientist or engineer designing a laser system for specific applications. For example, if an assembly-line laser welder outputs 10 kW of laser power continuously and has a *wall-plug efficiency* of 10% efficient, this suggests that there is a power need of 100 kW from the wall-plug. This means the system wastes 90 kW somewhere and that power will become heat that must be dealt with in some fashion.

The control subsystem typically consists of computers, sensors, control circuits, and input consoles. These components are usually straightforward and can be accounted for typically from the data supplied from the vendor about each. An example of such a component is a computer, which will have power consumption information printed on the back of it.

### 6.6.2 Power Conversion and Coupling Efficiencies

The laser assembly subsystem will have most of the losses that we must pay very close attention to. The first step will be taking power from the wall-plug and then conditioning that power to be of the type needed by whatever the pump source is. This might require transformers, pulse-forming circuits, and other complex electrical systems that will inherently have power losses with each conversion from one power form to another. The power conditioning subsystem will have an efficiency associated with it we will define as $\eta_{\text{Pconv}}$.

Once the power is converted to a form the pump source can use, it must be coupled into the pump system. The power coupling from the power conditioning subsystem to the pump source is typically done through some type of electrical connectivity, which will vary based on the power format. For high-voltage pulsed system low-loss, low-inductance high-voltage cables are required. For CW lasers the power is typically direct current high voltage so the cables must be low loss and low capacitance. Whatever the system is, there will also be some losses in the process of coupling electrical power from the power conditioning subsystem to the pump source. We will define the efficiency of coupling power here as $\eta_{\text{Pcoup}}$.

At this point, we have power delivered to the pump source and the pump source must convert the electrical power to a form that can be absorbed by the *active medium*. This might be done by dumping all the power into a flash-lamp and creating an optical flash of light. The power could be used to drive laser diodes to pump the *active medium*. Like in the case of the HeNe the power might be delivered directly into the medium gas to create a plasma discharge. Whatever the pumping mechanism may be, there will be an efficiency of delivering that pump power into the gain medium. In the case of the flashlamp some of the light from the optical flash will be reflected off the surface of the active medium and will be lost. Some of the light might be absorbed by impurities in the host material and converted to heat. In the case of a gas laser, some of the energy is lost in ionizing the gas molecules and again some is lost in unwanted transitions that are de-excited through transferring heat to the walls of the container. Again, there is an efficiency of supplying pump power into the active medium we will call $\eta_{\text{pump}}$.

### 6.6.3 Quantum Efficiency, Quantum Defect, and Slope Efficiency

Once the pump power is actually inside the *active medium*, it must be converted to laser energy through whatever process the specific type of laser implements. The laser power output divided by the power into the gain medium is called the *slope efficiency*, $\eta_{\text{slope}}$. Figure 6.25 shows a graph of (an arbitrary) laser output as a function of pump power input. The slope of the line in the graph is the *slope efficiency*. Typically, the graph is linear but in some cases it might not be. The *slope efficiency* can be determined by

$$\eta_{\text{slope}} = q_{\text{defect}}\eta_{\text{quantum}} \tag{6.30}$$

where $q_{\text{defect}}$ is the *quantum defect* which is the energy difference between the shorter wavelength pump photons and the longer wavelength laser photons. It is calculated by

$$q_{\text{defect}} = \frac{hc}{\lambda_{\text{pump}}} - \frac{hc}{\lambda_{\text{laser}}}. \tag{6.31}$$

The quantity $\eta_{\text{quantum}}$ is the *quantum efficiency* and is defined as the ratio of the number of photons of pump light that triggered stimulated emission divided

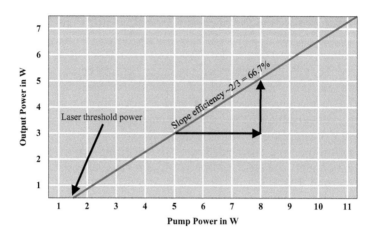

**FIGURE 6.25**
The slope efficiency is the ratio of output laser power to input pump power.

by the total number of input photons. This quantity is very close to 100% for most lasers but can be lower in some cases.

### 6.6.4 Cavity Efficiency

The last loss mechanism in the process going from wall-plug to laser light out of the system has to do with optical inefficiencies within the laser cavity itself. There will be some small losses within the optical components as well as within the gain medium based on how optically pure the material is and how well it can dissipate waste heat. This is the *cavity efficiency*, $\eta_{\text{cavity}}$, and will be specific for each laser.

### 6.6.5 The Laser Efficiency Calculation and Measurement

Now that we have addressed the losses between the wall-plug and the laser light output, we can revisit Equation 6.29 taking into account the other efficiencies defined between Sections 6.6.2 and 6.6.4 and develop an overall laser output equation based on efficiencies

$$P_{\text{laserout}} = \eta_{\text{cavity}} q_{\text{defect}} \eta_{\text{quantum}} \eta_{\text{pump}} \eta_{\text{Pcoup}} \eta_{\text{Pconv}} \left( P_{\text{in}} - P_{\text{thermal}} - P_{\text{control}} \right). \quad (6.32)$$

We will call Equation 6.32 the *laser output efficiency equation*. We can use information from the laser design process in the calculation or we can use measurements to determine the specific efficiencies within it depending on our goals. We should point out another aspect of making measurements for this equation and that is the inherent error in any scientific measurement. With each of the measurements of each of the efficiencies, there will be an uncertainty value involved. If we are making PIB measurements and then comparing that to the wall-plug output to gather an overall wall-plug efficiency then we also must realize that our PIB detector will have some losses due to the wavelength response curve of the detector material, reflections off of the first surface (and others), and even quantum efficiencies in converting the laser light to electrically measured signals. Therefore, our measurement device will have an efficiency of measurement, $\eta_{measure}$, that must be accounted for in the process. Most commercial detectors will have this information available and the electronics giving the measurement output data will be calibrated for such. In some cases, the laser scientist or engineer will have to build their own measurement device and these errors, uncertainties, and efficiencies will have to be accounted for.

## 6.7 Chapter Summary

In this chapter, we learned a lot of very practical and useful information that every laser scientist or engineer will use at some point during their career. We talked mainly within this chapter about descriptions, characteristics, and parameters that tell us specifics about specific lasers.

In Section 6.1, we learned the differences between CW and pulsed laser systems. We discussed the various methods of generating laser pulses through *gain switching* and we also learned three methods for measuring the *temporal profile* of those pulses. Then in Section 6.2, we learned about the various types of modes that can exist within a laser. We discussed measuring these *temporal* and *spatial modes* and what they tell us about the laser output beam.

In Section 6.3, we learned about the spectral content of laser beams and that we can use both the *temporal profile* and interference patterns to determine the *linewidth*. We discussed the optical device called the *etalon* and how to use it to make actual *linewidth* measurements of real laser beams. We also discussed various laser cavity configurations that can be used to tune laser output based on adding dispersive elements like prisms into the cavity.

In Section 6.4, we learned that laser beams do not truly go on and on perfectly *collimated* forever because of diffraction. We learned about *beam divergence* and how to use optical components to create a *beam collimator, expander*, or *reducer* to correct the diverging beam. From that discussion, we moved into more details about the laser beam in Section 6.5 where we learned about the power within a beam and how to describe it. We also learned that there are parameters used to describe how good a laser beam is through a concept called *beam quality*. We also discussed how to make power measurements of laser beams.

Then finally in Section 6.6, we went into great detail in discussing how a complete laser system converts power from the wall-plug into laser output light. We developed the *laser output efficiency equation* that will enable the laser scientist or engineer to make design choices when building a laser or to understand details about a laser system through various measurements.

Through this chapter, we have begun to not only learn what lasers are, but also a little bit about how to describe and use them. This chapter has information throughout it that is of the pragmatic type to be implemented in real-world situations. The information discussed here has proven important and useful to the author on many occasions during the design, research, the laboratory testing, and final application steps of many laser systems.

## 6.8 Questions and Problems

1. Describe the difference between CW and pulsed laser beams.
2. What is *gain switching*?
3. What does FWHM mean?
4. What are three techniques for measuring the *temporal profile* of a laser beam?
5. What is *pulse width*?
6. What is a *longitudinal mode*?
7. What is a *transverse mode*?
8. How many longitudinal modes can operate in a cavity 1 m long at a wavelength of 532 nm?
9. A laser cavity with wavelength 810 nm has a *mode number* of 100. How long is the cavity?
10. What is the mode spacing in frequency for a solid-state laser with a laser rod index of refraction of 1.801 and length 15 cm?
11. If the laser wavelength in Problem #10 is 1,064 nm, what is the mode spacing in wavelength?
12. A laser with a wavelength of 632.8 nm and a linewidth of 0.0005 nm operates in a gas discharge with index of refraction of 1.021. What is the *coherence length* of this laser?
13. A laser has a coherence length of about 75 cm operating at a wavelength of 480 nm. What is the linewidth in both frequency and wavelength?
14. Draw a $TEM_{13}$ rectangular spatial profile.
15. What does TEM mean?
16. Draw a $TEM_{22}$ cylindrical spatial profile.
17. What is the difference between a $TEM_{00}$ rectangular and cylindrical spatial profile?

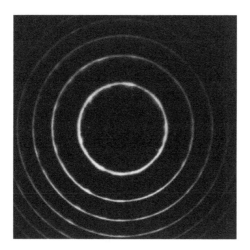

**FIGURE 6.26**
Interference pattern from 590-nm dye laser pulse after passing through an etalon with a FSR of 1,500 MHz. (Courtesy U.S. Army.)

18. An *etalon* has equal mirror reflectance for each surface of $R = 0.97$. What is the *finesse* of the *etalon*?

19. A pulsed dye laser with Rhodamine 6G in methanol operating at 590 nm with a multi-prism Littrow grazing incidence grating as the back-reflector was tuned for narrow *linewidth*. Use the interference pattern acquired with an *etalon* with an FSR = 1,500 MHz shown in Figure 6.26 to determine how narrow the *linewidth* of the laser was.

20. What was the *finesse* of the *etalon*?

21. The beam of a HeNe is 5 mm in diameter at the exit of the laser. At 1 km distance, the beam is 1 m in diameter. What is the beam's *divergence*?

22. A 532-nm laser beam is collimated through a telescope with the aperture diameter of the primary optic of 50 cm. How big is the central order spot at 5 km?

23. The same laser in Problem #22 is projected at the Moon (assume 400,000 km away). What is the spot diameter on the Moon?

24. If the same laser from Problem #23 has an output power of 1 kW, what is the irradiance of the spot on the Moon?

25. What is PRF?

26. Using the temporal profile from Figure 6.27 determine the PRF of the laser.

27. What is the *instantaneous power* for the laser in Problem #26?

28. Assuming the pulse train continues for the laser in Problems #26 and #27 what is the *average power*?

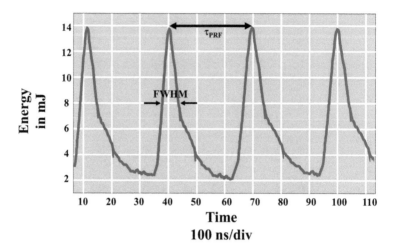

**FIGURE 6.27**
Temporal profile of laser for Problem 25.

29. What is the duty cycle of the laser in Problems #26–28?

30. What is PIB?

31. What is BPP?

32. A laser produces a beam with BPP = $\lambda/\pi$ what is the *beam quality*?

33. A laser gets 1 MW power from a wall-plug and produces a 100 kW beam output power. What is the wall-plug efficiency of the laser?

34. A laser gain medium uses a 532-nm pump source and generates 590-nm laser light from stimulated emission. What is the *quantum defect* of the medium?

35. If the *quantum efficiency* of the laser in Problem #34 is 97% what is the *slope efficiency* for the laser?

36. Simulation Project: Use the *laser output efficiency equation* given in Equation 6.32 in a computer program and simulate various efficiency ranges based on different assumptions about each of the components in Figure 6.24.

# 7

## How Do We Use Lasers Safely?

Up to this point, we have learned about what lasers are and how to describe them. We should have enough of a basic understanding with the information studied thus far to begin using lasers in the laboratory research and for other applications. In many cases, the novice laser scientist or engineer will have a more senior scientist or engineer to observe and learn from, but in some cases the novice might be on his or her own.

Lasers are a lot like automobiles. They are very useful and utilitarian and once we learn how to use them safely, they open up many aspects of the world that were unattainable before. They are also extremely dangerous if improperly handled. It is a very good idea for the first thing young teens are shown in driver's education classes is a video of horrific accidents and the outcome and consequences of using cars improperly. Along the same lines, the author suggests that all laser scientists and engineers from time to time use their favorite search engine and search the Internet for "laser eye damage," "laser injuries," and "laser accidents." This is not to simply scare us away from using lasers. On the contrary, seeing what lasers can do to the human body when used unsafely gives us respect for them and for the safety rules. And just like the automobile, once we learn to use them safely an entirely new and exciting world is opened up for us.

In this chapter, we will talk a little about laser safety calculations and laser safety rules that are typically covered in laser safety courses. But what isn't usually covered in those courses is practical experience about laser safety that is usually best handed down from one laser scientist or engineer to the next. We will discuss some of these ideas, practices, and protocols to help the novice begin to use lasers safely.

## 7.1 The Laser Safety Basics

### 7.1.1 ANSI Z136.1

Before starting on any laser project, the very first thing the laser scientist or engineer **MUST** do is to acquire a copy of the *American National Standard for Safe Use of Lasers, ANSI Z136.1* publication that is approved by the American National Standard Institute and is maintained by the Laser Institute of America (see Figure 7.1).

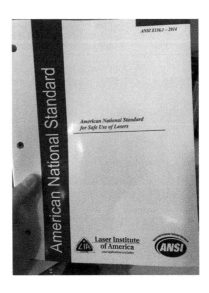

**FIGURE 7.1**
Before starting any laser project, the laser scientist or engineer must acquire a copy of this publication.

The foreword to the book says,

> Since 1985, Z136 standards have been developed by the ANSI Accredited Standards Committee (ASC) Z136 for Safe Use of Lasers. A copy of the procedures for development of these standards can be obtained from the secretariat, Laser Institute of America, 13501 Ingenuity Drive, Suite 128, Orlando, Fl 32826, or viewed at www.z136.org.
>
> The present scope of ASC Z136 is to protect against hazards associated with the use of lasers and optically radiating diodes.

It is recommended that all laser scientists and engineers refer to the *ANSI Z136.1* often and keep it handy as a desk reference. There are courses and online information based on this publication as well that are quite useful in learning about the safe use of lasers.

### 7.1.2 Laser Classes

The *ANSI Z136.1* defines several *classes* of lasers based on output power and potential safety hazards. Knowing what each of these classes are, is extremely important in that different safety protocols are required for each class of lasers.

#### 7.1.2.1 Class 1 Laser System

This laser system is supposed to be incapable of causing damage of any sort to the human body with direct exposure and requires no special control measures or protocols for safety. Figure 7.2 shows a typical Class 1 laser safety sticker.

### 7.1.2.2 Class 1M Laser System

This laser system is safe for all conditions except when viewing the beam through magnifying optics that increase the irradiance of the beam. In other words, the beam coming from the laser is too big in diameter (meaning fewer watts per meter square) to cause damage to the eye, but if it is reduced or focused to the point where the irradiance is high enough to cause damage then this laser becomes a Class 1M. Figure 7.3 shows a typical Class 1M laser safety sticker.

### 7.1.2.3 Class 2 Laser System

This class of laser is any laser emitting light in the visible range between 400 and 700 nm and does not emit enough light to cause damage faster than the blink response of the eye, which is about 0.25 s. If the sun were a laser, it would be of this class as staring at the sun much longer than the blink response time can cause damage to the eye. Figure 7.4 shows a typical Class 2 laser safety sticker.

### 7.1.2.4 Class 2M Laser System

Much like Class 1M this laser system is safe because the beam is too large and has too small an irradiance to cause damage (in this case faster than

```
CLASS 1 LASER PRODUCT
```

**FIGURE 7.2**
A typical Class 1 laser sticker.

**FIGURE 7.3**
A typical Class 1M laser sticker.

**FIGURE 7.4**
A typical Class 2 laser sticker.

**FIGURE 7.5**
A typical Class 2M laser sticker.

the blink response) without other optical components reducing the beam. But if it can cause damage with reducing optics within the blink response time then it is Class 2M. Figure 7.5 shows a typical Class 2M laser safety sticker.

### 7.1.2.5 Class 3R Laser System

This class of laser is typically less than 5-mW continuous wave (CW) in the visible and while it can cause an eye injury, the probability is low. For invisible wavelengths (such as IR and UV) and for pulsed lasers the safety limits vary. It is safe to view this laser from a diffuse reflection (such as on a card, piece of paper, or the wall) and it does not represent a fire hazard. Figure 7.6 shows a typical Class 3R laser safety sticker.

### 7.1.2.6 Class 3B Laser System

This class of laser may be hazardous with direct and specular reflection viewing of the beam. Diffuse reflections are not typically hazardous. This class of laser is also not typically a fire hazard. Figure 7.7 shows a typical Class 3B laser safety sticker.

**FIGURE 7.6**
A typical Class 3R laser sticker.

**FIGURE 7.7**
A typical Class 3B laser sticker.

### 7.1.2.7 Class 4 Laser System

This class of laser is a hazard to the eye or the skin from direct beam exposure. It may be a hazard to eye and skin from specular reflections and may be a hazard to the eye with diffuse reflection viewing. It may also be a fire hazard. This type of laser may also produce *laser-generated air contaminants* and hazardous plasma radiation. Figure 7.8 shows a typical Class 4 laser safety sticker.

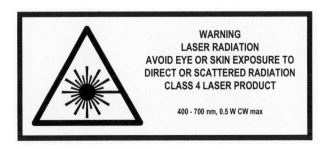

**FIGURE 7.8**
A typical Class 4 laser sticker.

### 7.1.3 Accessible Emission Limit and Maximum Permissible Exposure

The *ANSI Z136.1* defines the limiting output of lasers for each laser class. This limit is called the *accessible emission limit* or AEL. The AEL is defined as

$$AEL = MPE \times LA \tag{7.1}$$

where MPE is the maximum permissible exposure and LA is the *limiting aperture* area (typically the eye with an aperture diameter of 7 mm). MPE is the level of laser output (might be pulsed or CW) that a person can be exposed to without any harm coming to the eyes or skin. We can rewrite Equation 7.1 as

$$AEL = MPE \times \pi r_{eye}^2. \tag{7.2}$$

We should also adjust Equation 7.2 based on the MPE for eye or skin with subscripts as

$$AEL = MPE_{eye} \times \pi r_{eye}^2 \tag{7.3}$$

and

$$AEL = MPE_{skin} \times \pi r_{eye}^2. \tag{7.4}$$

**Example 7.1: AEL of the HeNe Laser**

We should note here that the MPE might be for the blink response time of 0.25 s or it can be for staring of 10–30,000 s. As an example, the MPE for blink response eye exposure for a HeNe laser at 632.8 nm is 2.6 mW/cm² and for staring is 1 mW/cm². Therefore, the respective AELs are

$$AEL_{blink} = 2.6 \frac{mW}{cm^2} \times \pi (0.7 \, cm)^2 = 4 \, mW \tag{7.5}$$

and

$$AEL_{stare} = 1 \frac{mW}{cm^2} \times \pi (0.7 \, cm)^2 = 1.54 \, mW. \tag{7.6}$$

### 7.1.4 Nominal Ocular Hazard Distance and Nominal Skin Hazard Distance

In order to determine if a laser is safe we need to determine if it is dangerous just outside the aperture or if it is only safe at some distance from that aperture. The minimum safe distance for the eye is called the *nominal ocular hazard distance* or NOHD and is found by

$$NOHD_{blink\,or\,stare(eye)} = \frac{1}{\Theta} \sqrt{\frac{1.27\Phi}{MPE_{blink\,or\,stare(eye)}}} \tag{7.7}$$

where $\Theta$ is the beam divergence in radians as defined in Equation 6.17 and $\Phi$ is the laser power.

The minimum safe distance for skin is the *nominal skin hazard distance* or NSHD and is found by

$$\text{NSHD}_{\text{short or long(skin)}} = \frac{1}{\Theta}\sqrt{\frac{1.27\Phi}{\text{MPE}_{\text{short or long(skin)}}}}. \tag{7.8}$$

It is a little funny sounding to use subscripts of *blink* or *stare* for NSHD as the skin cannot blink or stare so we change the subscripts to *short* or *long*, but the same exposure times of 0.25 s for *short* exposures and 10–30,000 s for *long* exposures are used.

### 7.1.5 Nominal Hazard Zone

Now that we know how to calculate the NOHD and NSHD for various lasers, we can determine the three-dimensional volume around the laser where danger from the beam exists. This volume is called the *nominal hazard zone* or NHZ. The NHZ is usually an overlay on a map of the laser lab or experiment range showing where dangers exist. The NHZ can be determined for direct beam exposure, specular and diffuse reflections, and for skin hazards.

The NHZ is found by

$$\text{NHZ}_{\text{short or long(eye or skin)}} = \frac{1}{\Theta}\left(\sqrt{\frac{1.27\Phi}{\text{MPE}_{\text{short or long(eye or skin)}}}} - a\right) \tag{7.9}$$

where $a$ is the diameter of the laser aperture in centimeters. The NHZ is the distance from the laser source where there is still danger. It may vary with angle so the hazard volume mapped out might not be a perfect spherical volume and is typically a three-dimensional blob. In fact, NHZ is what most laser scientists and engineers think of when speaking of this danger blob or zone, but the NHZ is a single distance based on specific parameters and really should be called the NHD or nominal hazard distance (but we won't break from tradition here). A better description of this blob would be to think of a *nominal hazard volume* or NHV, which is a function of the NHZ at specific spherical coordinates $\theta$ and $\phi$ where the NHZ is actually $r$. In other words,

$$\text{NHV}(r,\theta,\phi) = \frac{1}{\Theta}\left(\sqrt{\frac{1.27\Phi}{\text{MPE}}} - a\right)C_{nm}\widehat{\theta_n}\,\widehat{\phi_m} \tag{7.10}$$

where $C_{nm}$ is a correction factor that might even be a complex function based on the viewing angle, specular or diffuse reflections, and even atmospheric distortions, and $\theta_n$ and $\phi_m$ are angles for specific points on the surface of the blob. This blob is best calculated through a computer analysis or simulation that will build up the three-dimensional volume. The laser scientist or engineer can do a few values by hand if necessary but there are modern software packages that will do this calculation.

## 7.1.6 Threshold Limit and Optical Density

The *threshold limit* is the maximum irradiance allowed before damage to the eye or skin (or sometimes sensor or other device) will occur. In order to protect the eye, skin, or device a filter consisting of a specific *optical density* (OD) is used. OD is simply a logarithmic attenuation factor where an OD of 1 is an attenuation factor of 10. An OD of 6 would be an attenuation factor of 1,000,000. Figure 7.9 shows a graph of the OD required to attenuate visible laser beam irradiance output to safe viewing levels at the output aperture of the laser for less than 10 s duration. The OD required for safety is calculated as

$$OD = \log_{10}\left(\frac{H_p}{MPE}\right) \tag{7.11}$$

where $H_p$ is the output of the laser in the same units as the MPE for the specific calculation.

## 7.1.7 The Laser Safety Officer and the Standard Operating Procedure

Before embarking on any laser experimentation or applications in the lab or on a test range there must be a laser scientist or engineer with appropriate training acting as the *laser safety officer* or LSO. There are many online

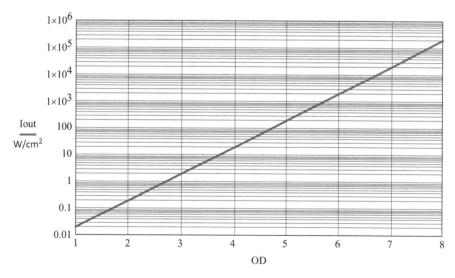

**FIGURE 7.9**
OD required to attenuate a visible CW laser beam for safe viewing for less than 10 s.

and short courses where a scientist or engineer can be trained in detail on laser safety and actually become certified as an LSO approved by certain professional communities. Certified or not, all laser scientists and engineers must absolutely adhere to laser safety processes and procedures in order to maintain a safe environment for themselves and their coworkers.

Typically, a standard operating procedure or SOP is generated for use of any laser device within the lab. That SOP will point out any potential hazards as well as required safety measures. All personnel in the laser lab or on the range must familiarize themselves with the SOP and any other safety requirements for their own safety. In the end, we are all responsible for our own safety and should follow laser safety protocols wisely.

## 7.2 Types of Injuries Possible from Laser Beams

### 7.2.1 Visual

Figure 7.10 shows the external features of the human eye. Depending on the type of laser beam, any or all of these anatomical features can be at risk from exposure. The eye will respond to exposure and both blinking the *eyelids* and tear production can occur. The *iris* will also reduce but typically not fast enough to protect the internal features of the eye from exposure. The *pupil* is where the light enters and passes into the internal portion of the eye.

Figure 7.11 shows a very detailed three-dimensional anatomical drawing of the human eye with most of the main features labeled. The optical pathway through the eye is where laser injuries are likely to occur. The *macula* (or *macula lutea*) area labeled number 25 is where the *lens* will focus light and

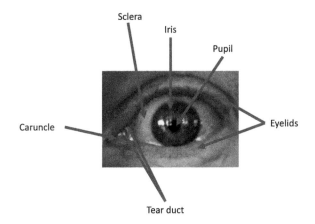

**FIGURE 7.10**
The external components of the eye.

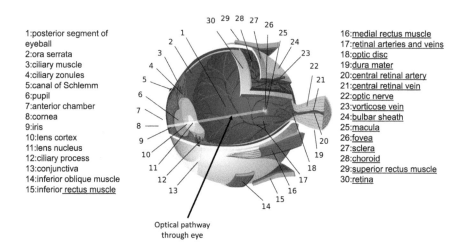

1:posterior segment of eyeball
2:ora serrata
3:ciliary muscle
4:ciliary zonules
5:canal of Schlemm
6:pupil
7:anterior chamber
8:cornea
9:iris
10:lens cortex
11:lens nucleus
12:ciliary process
13:conjunctiva
14:inferior oblique muscle
15:inferior rectus muscle

16:medial rectus muscle
17:retinal arteries and veins
18:optic disc
19:dura mater
20:central retinal artery
21:central retinal vein
22:optic nerve
23:vorticose vein
24:bulbar sheath
25:macula
26:fovea
27:sclera
28:choroid
29:superior rectus muscle
30:retina

Optical pathway through eye

**FIGURE 7.11**
Three-dimensional view of the eye with anatomical features labeled. (Courtesy GNU Free Documentation License, Version 1.2 by Chabacano.)

where the irradiance is likely to be highest. It is near the center of the *retina* (number 30), is about 5.5 mm or so in diameter, and is responsible for high resolution and color vision in normal lighting situations. Within the *macula* is the *fovea* (number 26) which is an area of highly packed visual receptors called *cones*. It is the fovea where the highest-resolution imaging is done by the eye. If the *macula* is damaged vision will be impaired.

Figure 7.12 is a cross-sectional view of the eye that gives a simpler and easier to understand depiction. Figure 7.13 takes that cross-sectional view

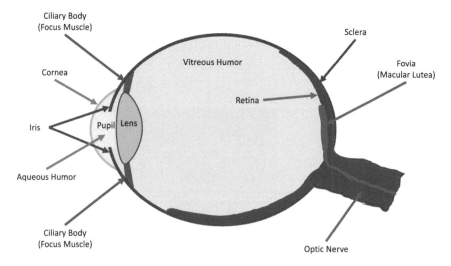

**FIGURE 7.12**
Cross-sectional view of the eye with anatomical features labeled.

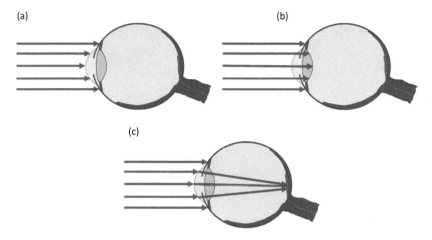

**FIGURE 7.13**
The transmission of laser beams through the eye: (a) far ultraviolet and far infrared, (b) near ultraviolet, and (c) visible and near infrared.

and shows the transmission of laser light at various regions of the optical spectrum. Figure 7.13a shows that incident far-ultraviolet and far-infrared laser lights are absorbed on the surface of the eye and any damaged caused by lasers at those wavelengths would occur there. Figure 7.13b shows that near-ultraviolet light passes through the surface of the eye into the lens region. Figure 7.13c shows the transmission of visible and near infrared which gets focused by the lens of the eye onto the retina.

At the focus the irradiance is highest and where damage occurs. Unfortunately, the focus of the eye is also where the most important vision receptors are (the *fovea* of the *macular region*) and where the worst impact to vision can occur. The laser scientist or engineer must be vigilant in safety as once the retina is damaged there is little that can be done to repair it. A poor decision in safety might only take a fraction of a second to cause permanent reduction or even total loss of vision. Therefore, it is extremely important to follow the manufacturer's recommendations for laser safety with all lasers and to wear the appropriate recommended eye protection.

### 7.2.2 Skin

Laser damage to the skin is a more obvious scenario than eye damage. Exposure to a high-power laser beam of about 1 W or more can cause burns. At levels of under about 5 W, the human flinch reaction will protect from serious burns—just like touching a hot plate cause us to reflexively jerk our hands back. At powers higher than 5 W, the burns can occur even within the flinch reaction time.

Figure 7.14 shows a detailed diagram of the human skin. Depending on the wavelength, the surface tissue of the skin can be damaged or the laser

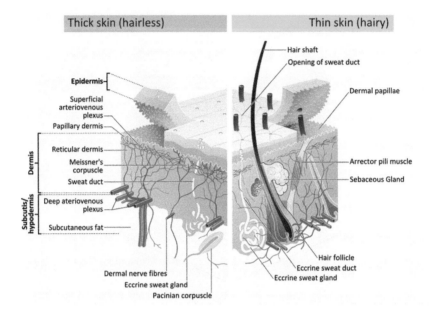

**FIGURE 7.14**
A cross-sectional view of the human skin. (Courtesy Creative Commons Attribution-Share Alike 3.0 Unported by Madhero88 and M. Komorniczak.)

light can penetrate as far as the subcutaneous tissue. Certain wavelengths are absorbed easily by hair and will superheat the follicle and kill it, which is how laser hair removal works. Some skin pigmentations absorb certain wavelengths easily and that is how blemishes, birthmarks, and some tattoo inks can be removed.

Figure 7.15 shows the skin diagram (shown in Figure 7.14) with burn degree based on burn penetration depth. As the laser beam burns deeper into the skin, the injury becomes more and more serious. First-degree burns are fairly superficial and heal fairly quickly—similar to sunburns. As the degree of the burn progress from second to fourth, irreparable damage can occur.

## 7.3 Other Safety Considerations with Laser Systems

### 7.3.1 Electrical Hazards

Many laser systems require high-voltage systems in order to generate enough optical power either to ionize a gas or to pulse a flashlamp or discharge rapidly enough to cause the laser population inversion. In many systems sometimes much greater than 10,000 V is required and the amperages might

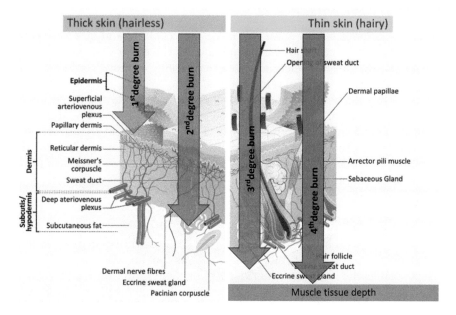

**FIGURE 7.15**

A cross-sectional view of the human skin with burn degree penetration depth shown. (Skin image courtesy Creative Commons Attribution-Share Alike 3.0 Unported by Madhero88 and M. Komorniczak.)

be even greater than 10 A. The power system is very specific to the type of laser and the output powers required.

Due to the need for such high powers, the need for high-voltage and high-current electrical safety protocols is often a necessity. All laser scientists and engineers must familiarize themselves with the laser systems being used and the electrical power safety requirements.

### 7.3.2 Fire Hazards

Many lasers implement high-voltage discharges, spark gaps, flammable liquids and other materials, and produce optical light that might be bright enough to cause fires. The laser scientists and engineers using any laser system must know all of the potential fire hazards present and must have protocols in place to prevent them as well as to extinguish them if they were to occur.

Consider a high-power pulsed dye laser using methanol as the solvent. If for whatever reason the laser were to catch fire there is still likely to be very high voltages present and certain fire extinguishing techniques could pose an electrical shock hazard to the fire fighter. The appropriate fire mitigation strategies for each laser system must be in place and operators must be trained in them before ever turning on the laser.

### 7.3.3 Asphyxiation Hazards

Many lasers use gases for gain media, drying, cooling, and other purposes and most of these gases are not breathable. In the event that there is a gas flow loop leak, the laser scientist or engineers being exposed could suffocate or be affected with harmful ramifications.

Even if the gas being used seems as harmless as helium, it could pose a serious safety hazard if it leaks in large enough quantities. Breathing low levels of helium might be humorous, as it affects the vocal chords and will produce the "munchkin voice." But humans cannot survive on helium and breathing too much of it by accident could cause the loss of consciousness which could lead to much worse problems.

### 7.3.4 Toxic Hazards

Many lasers use highly toxic materials. Most of the laser dyes in dye lasers are extremely toxic and they typically use solvents that are also toxic. Chemical lasers such as the hydrogen fluoride (HF) or deuterium fluoride (DF) lasers use such toxic levels of fluorine that the seconds of exposure could literally vaporize the human lungs if breathed in. Serious safety protocols must be put into place for such laser systems and they must be strictly adhered to.

### 7.3.5 Explosive Hazards

Many lasers also pose potential explosive hazards due to either volatile chemicals, gas overpressures, high-voltage components, or a combination of these. There are even some lasers, which use explosives to trigger the population inversion. The laser scientist and engineer must be aware of any such potential explosive hazards and implement mitigation strategies.

## 7.4 Practical Safety Considerations

### 7.4.1 Specular Reflections from Common Attire and Objects

The most common injury to laser scientists and engineers is eye damage from stray laser beams, reflecting from a surface the user had not accounted for. These types of objects can be from clothing, jewelry, security badges, glasses, and pretty much any object that is shiny enough to produce a specular reflection.

Figure 7.16 shows a Class IIIb 650-nm diode laser used for demonstration of these types of specular reflections. Figure 7.17 shows stray specular reflections from a wedding ring. There are many very bright spots being reflected in multiple random directions that could pose threats to the eye. Figure 7.18

**FIGURE 7.16**
A Class IIIb laser used to demonstrate stray light safety.

**FIGURE 7.17**
Laser beam specular reflections from a wedding ring.

shows the laser reflection from: (a) the surface of a watch and (b) the metal components of the watch structure. There is significant light from each being reflected.

Figure 7.19 shows a reflection from a pair of reading glasses. A very significant amount of the light is bounced off the first surface of the glasses

(a)                                          (b)

**FIGURE 7.18**
Laser beam specular reflections from a watch: (a) directly from glass surface and (b) from metal components.

**FIGURE 7.19**
Laser beam specular reflections from a pair of reading glasses.

and could be a hazard to someone else in the room. Figure 7.20 shows a reflection off of a plastic security badge. In many labs, people will need badges to open doors or to gain access to them. These badges often pose a stray reflection hazard. Figure 7.21 shows stray reflections from a partially filled plastic water bottle. Again, these reflections could pose a serious safety hazard. Figure 7.22 shows the reflection of the laser off a pair of scissors. The metal blades of the scissors reflect a significant amount of the laser beam and could prove to be a hazard as well.

**FIGURE 7.20**
Laser beam specular reflections from a plastic security badge.

**FIGURE 7.21**
Laser beam specular reflections from a plastic water bottle.

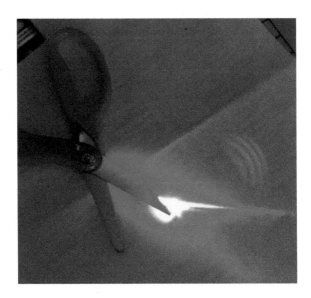

**FIGURE 7.22**
Laser beam specular reflections from a pair of scissors.

From a safety perspective, it is a very good practice to have a mental (if not a real one) checklist of things that is always done before ever turning on a laser. Always remove any jewelry, badges, and other shiny apparel. If clothing is shiny or has highly polished buttons, it is a good practice to have a lab coat or other covering material to prevent exposing these objects to the laser beam. Never set bottles, cups, cans, or other containers where they might be in the optical path. And be careful not to use shiny tools like scissors, wrenches, or screwdrivers near the laser beam or while it is on.

### 7.4.2 Even Diffuse Reflections Can Be Dangerous!

Figure 7.23 shows the laser beam incident on a piece of paper. The paper is a fairly diffuse reflector, but from the picture it is clear that there is a significant amount of laser energy being spread out. Diffuse reflections are not as dangerous as specular ones, but if the laser is powerful enough such reflections can cause eye damage from certain distances.

It is a very common practice that laser scientists or engineers use business cards (and index cards) to view a laser beam in the lab. Figure 7.24 shows a laser beam incident on a standard business card producing both a specular reflection and a diffuse reflection. Be aware of such items as they can cause a safety risk if we do not realize that they can redirect the laser light in such ways.

Figure 7.25 shows the laser beam incident on a fingernail. Believe it or not, the fingernail can be highly reflective with both diffuse and specular

**FIGURE 7.23**
Laser beam diffuse reflection from a piece of paper.

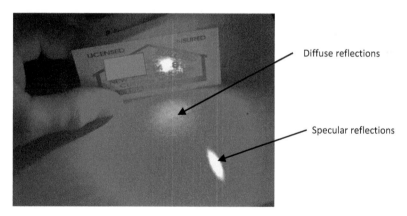

**FIGURE 7.24**
Laser beam diffuse and specular reflection from a business card.

components. Be careful placing your hands and fingers through any laser beam pathways.

### 7.4.3 Beam Stops

Stray reflections will occur in the lab wherever there is an optical component in the path of the laser beam. Figure 7.26 shows an example of optical

**FIGURE 7.25**
Laser beam diffuse and specular reflection from a fingernail.

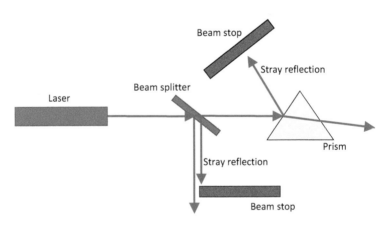

**FIGURE 7.26**
Use beam stops for stray reflection safety.

component setup where first-surface and even second-surface reflections might cause stray reflections that could be a safety hazard. It is a good idea plan for such reflections and look for them before becoming engrossed or absorbed in the experiment work. Once these reflections are identified, a beam stop should be implemented to prevent the reflection from traveling to unwanted regions of the lab. Beam stops can be made of paper, cardboard, wood, metal, or other materials depending on the power and wavelength of the laser beam. They should be of a material that will absorb as much of the light as possible and have no specular reflection.

### 7.4.4 Apparatus Height and Stray Light

As an experimental laser apparatus is designed, constructed, and implemented, the operator should consider the height as a safety factor. As most experiments are constructed on an optics table, it is typically preferential to have all the light pathways well below eye level. Any stray light that travels out of the plane of the experiment must be accounted for and blocked to prevent it from reaching eye level and posing a hazard. It is also a good practice to close or protect the eyes when bending down to perform tasks near the floor (such as picking up a tool or tying a shoe) because the eyes will likely pass through the beam plane on the way down.

### 7.4.5 Invisible Beams

Many lasers produce invisible beams either in the ultraviolet or in the infrared. For ultraviolet laser beams typically a simple index card can be used to find the beam as it will cause the card to fluoresce. Other cards with dyes that fluoresce in the visible when exposed to ultraviolet light can also be used. For infrared beams sometimes thermal paper can be used or camera systems that can detect infrared light. Invisible beams are more hazardous than visible laser beams in that it is not obvious where the stray light reflections are. It is a wise protocol to operate the laser at safe low powers first while searching for stray light and doing any alignment procedures.

### 7.4.6 Discharging and Disconnecting the Power

When disconnecting power components of many lasers there might be power-storing capacitors holding very high electrical charges even though the system is unplugged from external power outlets or sources. Before doing maintenance or other activities with lasers that use high voltage, make certain that all power-storing devices have been discharged completely. A standard way of doing this is to use a discharge-grounding bar or cable that can be placed across both leads of the capacitor or storage network. Typically, a loud electric arc can be seen and heard during the discharging process. The grounding tool should be used several times to be certain no relaxation charge builds back up in the storage network. It is also a good idea to keep a grounding cable shorting out the network while performing work on the system and not removing it until the work is completed.

### 7.4.7 High Voltage, Don't Point Your Fingers!

One particular safety tip that, unfortunately, many laser scientists and engineers learn the hard way is that high voltage likes to collect and arc from and to points. It is human nature to point at things as we talk about them. This can be hazardous around high-voltage pulse forming networks, arcs, flashlamps, and other similar systems. Pointing a finger at such a device might create a pathway that electrical current can leap to, which in turn will cause severe electrical shock. So, don't point your fingers at high-voltage components.

### 7.4.8 Fires Can Be Invisible!

Many lasers use chemicals and solutions that are highly flammable. Many of these materials will burn and some of them are invisible when they burn. The fires can be in the infrared or ultraviolet and can still cause severe burns. Be aware of unusual heat signatures even though you might not actually see a fire—there could be a fire you just do not see. Dye lasers are a perfect example of where this can happen. Many dyes use the solvent methanol which when burning in daylight cannot be seen and is very dangerous.

## 7.5 Chapter Summary

Chapter 7 was all about using lasers safely. In many cases, the laser scientist or engineer has to know about how each classes of lasers work, the physics of population inversions, and rate equations and laser cavities, but oftentimes the safe use of the devices is overlooked. The goal here was to give the reader a brief overview of some of the key aspects of using lasers safely, and where to find more detailed reference material and training.

In Section 7.1, we learned about the laser safety basics and the ANSI's standard reference material *ANSI Z136.1*. It is this standard that is the starting point for any laser safety protocols to be implemented. Every laser lab should have a copy of this book available for all personnel. It is recommended that coursework, short courses, online courses, or other similar means be implemented to familiarize all personnel with the material before being allowed to perform laser experimentation.

In Section 7.2, we discussed the anatomy of the eye and skin and how lasers can be a hazard to both. Various types of eye and skin injuries were discussed. Safety concerns not caused by the laser light were then discussed in Section 7.3. And finally in Section 7.5, we discussed various techniques to mitigate some of the potential hazards from lasers and how to use them safely.

## 7.6 Questions and Problems

1. What is the ANSI standard for laser safety?
2. What class of laser is completely harmless to the human body?

3. What class of laser is unsafe when viewed through a magnification system?

4. What class of laser is safe within the blink response time of the eye?

5. What class of laser is unsafe if it can cause damage within the blink response time of the eye when viewed through magnifying optics?

6. What class of laser is typically less than 5 mW and can, though unlikely, cause damage to the eye?

7. What class of laser can cause damage to the eye from specular reflections?

8. What class of laser can cause damage to the eye from diffuse reflections?

9. Define NOHD.

10. Define NSHD.

11. A HeNe laser with an output energy of 10 W, a beam diameter at the exit aperture of 5 mm, and a divergence of 1 mrad is to be used on an open air range. What is the NOHD for this laser if the MPE for blink response is 2.6 mW/cm$^2$?

12. What is the NHZ?

13. What is confusing about the NHZ and how could it be better named?

14. What is the NHV?

15. Consider that there is about 1,555 W/m$^2$ of sunlight incident on the earth. From what we know about the MPE for the HeNe laser in Problem 11 calculate if it is safe to stare at the sun or is it only safe from the blink response?

16. What is OD?

17. Calculate the OD required to make the HeNe in Problem 11 safe to stare at right at the exit aperture.

18. Where is the most likely location for damage to occur in the eye if a Class IIIb visible laser beam is viewed directly?

19. What is the *fovea*?

20. Where will the damage from an ultraviolet laser beam likely occur within the eye?

21. Where will the damage from a far-infrared laser beam likely occur within the eye?

22. If a Class IV laser burns the skin all the way to the muscle tissue, what degree of burn would this be?

23. Name five non-laser light-oriented hazards that can be caused by a laser system.

24. Are earrings a potential safety risk in the laser lab?

25. What type of outer garment might be a good option when considering laser safety?

26. What should be done before performing maintenance on a high-power pulsed flashlamp-pumped laser power supply?

27. Why should we not point at high-voltage components?

28. When tying a loose shoelace in the laser lab while a laser is operating describe the process in detail to do this safely.

29. Why using a water-type fire extinguisher on a flashlamp-pumped Nd:YAG laser fire probably not a good idea?

30. Create a safety checklist of things to do before ever turning on a laser in the lab.

# 8

## What Are Some Laser Applications?

There are so many types of lasers available with different characteristics, output power, wavelength, beam shape, pulse widths, and so on that the number of applications for lasers is pretty much unlimited. In this chapter, we will discuss some of the more common uses of lasers, but again, we have to realize that in order to discuss them all would take a text in itself or maybe more.

In essence, the laser is simply a tool—a very versatile one. We will look at how optical experiments are configured and constructed in the laboratory using lasers as tools. As it turns out in most modern-day laboratories one type of laser is usually implemented when setting up another. Sometimes many lasers are needed to set up an experiment based on its complexity.

We will also look at higher-power laser systems and their uses for industry, defense, and other more exotic applications. The point of this chapter is not to go into details about every application for lasers. Instead, this chapter will discuss the more popular, pragmatic, and common uses that the beginning laser scientist or engineer is likely to come across very early in his/her career. For further details of specific applications typically more study and specialization will be needed; however, this chapter is a good starting point.

## 8.1 Lasers Used in Experiments

### 8.1.1 Alignment

As we studied in Section 6.2, many modes can exist within a laser cavity. However, in order for these modes to form, there must be a reflection path between two optical components. For this reflection path to exist the two optical components must be precisely parallel with each other to within fractions of a wavelength of the light being reflected. The best way to determine if two optical surfaces are parallel is to bounce laser beams off of them and to then adjust the relative position of the two optics until the reflected spots overlay upon each other.

Figure 8.1 shows a typical configuration for aligning a laser cavity. An alignment laser is used as the main tool for this purpose. Typical alignment lasers are helium–neon (HeNe), diode-pumped solid-state lasers, and sometimes other lasers might be used. The alignment laser should be Class 2 or

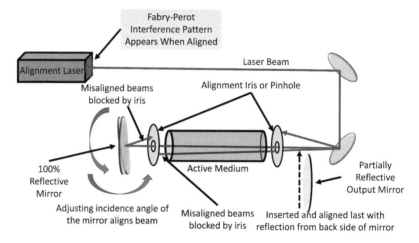

**FIGURE 8.1**
An alignment laser is used to set up a laser cavity.

less, so eye safety is not of major concern and it will also help us find where reflections and stray light might go once the higher-power laser is turned on. The alignment beam is best implemented when bounced off of two highly reflective (100%) mirrors that are adjustable in altitude and azimuth (two axes). These mirrors are used to place the beam through the center of the laser apparatus and active medium.

To more precisely align the system, we place an adjustable iris (or sometimes just a pinhole) with a smaller diameter then the active medium at the entrance and exit of the medium. In order for the alignment beam to make it through the smaller apertures the beam must be more precisely parallel with the axis of the laser-active medium and optical pathway. Alignment of the back reflector is typically done first. As the mirror is adjusted so that the reflected beam passes back through the irises and then back onto the exit aperture of the alignment laser, a set of interference rings will form on the face of the alignment laser. These rings will appear because the back reflector of the laser being aligned and the output coupler of the alignment laser form a Fabry–Perot interferometer once alignment has been accomplished.

Once the back reflector is aligned the output coupler is inserted into place. The mirror is adjusted until the reflection from the outside (back surface) of the output coupler is incident back onto the alignment laser-exit aperture. In some cases, the light from the back reflector is brighter than from the output coupler making it difficult to see if the output coupler is aligned or not. Placing an index card in front of the back reflector during alignment of the output coupler is typically a good practice. This will block the reflection from the back reflector and therefore the interference rings will vanish.

Once new interference rings appear from the cavity formed between the output coupler of the laser and the output coupler of the alignment laser the

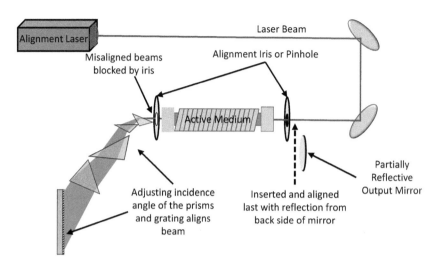

**FIGURE 8.2**
Alignment approach for a complex laser cavity.

card can be removed and the laser is then aligned. Also note, that before the higher-power laser is turned on the path to the alignment mirrors must be blocked, averted, or one of the mirrors must be removed or the alignment laser might be damaged.

Figure 8.2 shows a configuration for aligning a more complex optical cavity. The approach is very similar as described above with the added difficulty of aligning dispersive elements such as prisms and gratings. But in general, the approach is to create reflective pathways from the final reflective optic (in this case a grating) back through the cavity and onto the face of the alignment laser.

### 8.1.2 Probes

Lasers can also be used to probe into materials to determine characteristics of them. Tunable lasers, such as the dye laser, is very useful in probing the absorption, transmission, and fluorescence spectra of materials such as gasses, optically transmissive components, and even of surfaces.

Figure 8.3 shows a typical laser spectrometer experiment configuration. The tunable laser is incident on the subject material to be examined. The laser wavelength is tuned across the spectrum until the subject begins to fluoresce. The fluorescent light is captured by a telescope and directed into a spectrometer for analysis. The concept is quite useful for probing the absorption and fluorescence spectra of materials. This type of probe can also be useful for monitoring smokestacks of manufacturing plants for pollutants, detecting materials in locations that are difficult to place physical devices in,

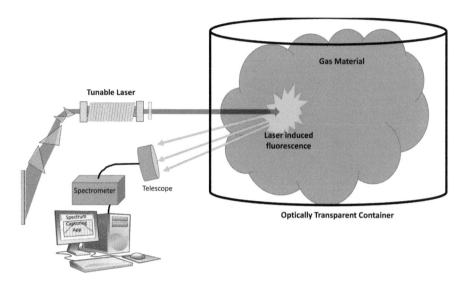

**FIGURE 8.3**
A typical laser spectrometer experiment.

remote detection or early warning of dangerous materials, and many other applications.

Figure 8.4 shows a depiction of the NASA Mars Curiosity Probe's laser-induced breakdown spectrometer known as the *ChemCam*. This probe uses a high-energy laser (HEL) pulse to ionize materials. As those materials are ionized, they fluoresce with a particular spectrum based on their composition. This device then enables the planetary scientist to study what the Martian soil and rocks are made of.

There are many types of laser probes. Some probe beams are used to determine the amplification saturation levels inside an active gain medium. Some laser probes are used to measure reflectance and/or absorption of surface materials. Some laser probes measure optical properties of materials such as optical density, birefringence, polarization, index of refraction, biological activity, electromagnetic properties, molecular and atomic properties, and many more. The potential for laser beams to be used to probe unknown parameters for scientific research is practically unlimited.

### 8.1.3 Precision Measurements

Due to the small wavelengths of laser beams, they are very useful for making precise measurements. One of the useful tools for precise measurements is the laser interferometer. There are many types of laser interferometers and specific configurations are suited for different applications.

The Michelson interferometer (shown in Figure 8.5) is one of the most widely used in laboratories and experiments throughout history and

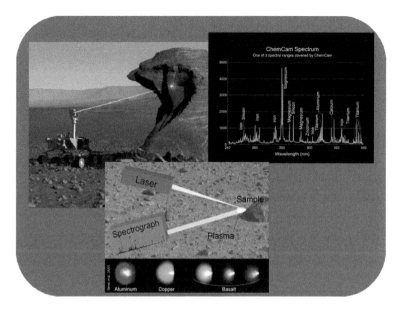

**FIGURE 8.4**
The laser-induced breakdown spectroscopy with ChemCam on the Mars Curiosity Probe can measure make-up of Martian rocks. (Courtesy NASA.)

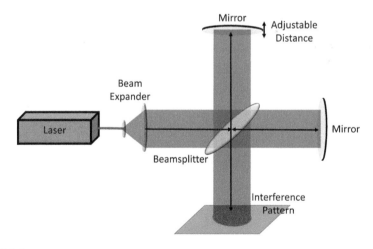

**FIGURE 8.5**
The Michelson laser interferometer is used to measure very small and precise path differences.

presently. This interferometer was invented in 1881 by Albert A. Michelson at the U.S. Naval Academy in Annapolis. Michelson used his experiment to make precise measurements of the speed of light in the classroom. At the time, Michelson used yellow light from a sodium flame, but in present days the light source is typically a narrow linewidth continuous wave laser.

As shown in Figure 8.5 there are two optical paths for the laser beam to travel. One path is fixed in distance and the other is adjustable. As the adjustable path distance is altered by even fractions of the laser wavelength the interference fringe pattern will change. Figure 8.6 shows the *Laser Interferometer Gravitational-Wave Observatory* or LIGO interferometer, which was designed to detect changes in path distance of 1/10,000th the width of a proton! The Hanford Site is in Richland, Washington, in the United States and was designed to detect gravitational waves caused by very dense and large masses moving in the space.

The irradiance of the fringes as a function of the distance difference, $d$, between the two optical paths is found by

$$I(d) = I_o \cos^2\left(\frac{2\pi}{\lambda}d\right). \tag{8.1}$$

Figure 8.7 shows a graph of Equation 8.1 with the wavelength assumed to be a HeNe laser beam at 632.8 nm. The graph shows that small distance differences of less than a micron will dramatically change the irradiance of the interference pattern. Thus, the laser interferometer is a tool that enables the laser scientist or engineer to make very precise measurements of very small distances or optical path changes.

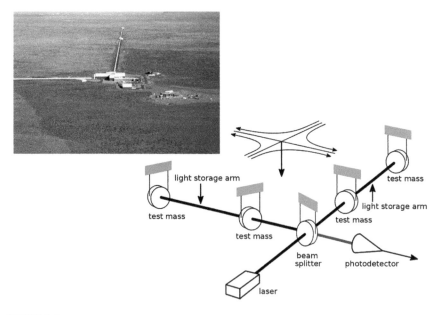

**FIGURE 8.6**
The LIGO interferometer at the Hanford Site in Richland, Washington, uses a Michelson interferometer to detect gravitational waves.

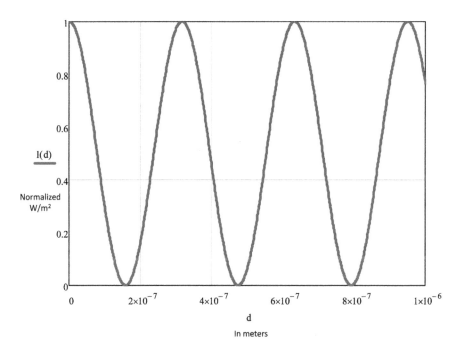

**FIGURE 8.7**
Irradiance of the Michelson interferometer as a function of the distance difference, *d*, between the two optical pathways (assumes HeNe wavelength of 632.8 nm).

## 8.2 Lasers in the Field

### 8.2.1 Laser Imaging, Detection, and Ranging

Since the invention of the laser in the 1960s, laser scientists and engineers have been using them to image, detect, and find ranges to objects. In fact, on Apollo 15 in 1971 astronauts used a laser system to map the surface of the Moon. This type of implementation is known as laser imaging, detection, and ranging or LIDAR (and like laser is now often written in the lower case as a word).

Lidar is implemented by using pulsed or modulated laser beams and laser detectors to determine precise distances to objects. Figure 8.8 shows a basic configuration for a lidar system. The laser pulses are split just outside the laser output coupler and one beam is directed to the receiving components (typically a telescope with photon counting detectors) while the other portion of the beam is directed onward to a distant object. Using the split beam as time reference the reflected beam's return time is then compared and the difference between them will be twice the travel time required for the

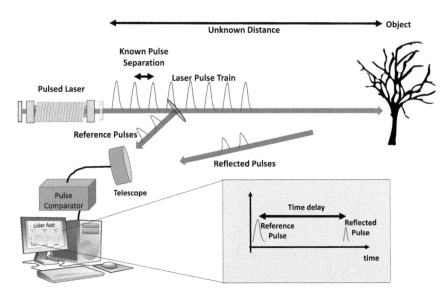

**FIGURE 8.8**
A lidar is used to detect and image objects at unknown distances.

light to travel from the laser to the object. This allows for the distance to be determined as

$$d = \frac{ct}{2}. \tag{8.2}$$

The lidar receiver will likely see returned light from particles in the air, multiple objects along the way, and any background (such as the ground or buildings). The various return times are used to map the various return distances to objects and therefore a three-dimensional map can be generated of the area being scanned. Figure 8.9 shows a lidar image of the Apollo Basin on the Moon taken from the Clementine LIDAR Altimeter instrument in 1994.

## 8.2.2 Leveling and Surveying

The laser is ideally suited for measuring surface heights for leveling and surveying purposes. The laser beam travels in a straight line and therefore can be used to place points accurately on a single plane. The laser level is very useful in construction applications as shown in Figure 8.10 where a laser level is being used to place tile accurately within the plane of a wall.

The same concept is useful for land surveying as well. Figure 8.11 shows a typical laser-surveying device. The typical laser-surveying device can be used to measure surface height as well as distances from a known or set geographical reference point.

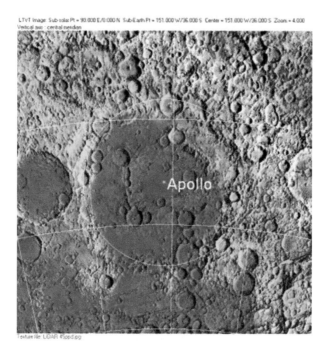

**FIGURE 8.9**
Lidar image of the Moon taken in 1994 by the Clementine spacecraft. (Courtesy NASA.)

**FIGURE 8.10**
The laser leveling tool is useful for placing points accurately on a plane. (Courtesy Creative Commons Attribution-Share Alike 3.0 Unported license image by Centurion.)

**FIGURE 8.11**
Typical laser surveying tool.

### 8.2.3 Laser Communications

Laser beams can be modulated just like radio and microwave beams can. Likewise, laser beams can be encoded with communications signals such as voice and data signals. Most modern-day Internet connectivity is accomplished via laser beams travelling through fiber optic cable. Free-space optical communications, often referred to as FSO, can also be accomplished with beams transmitting through the open air or even space.

Figure 8.12 shows a typical FSO laser transceiver. The device has multiple laser transmitters with small optical apertures in comparison to the larger receiver aperture in the middle. There is also an alignment monocular that aids the operator in pointing the system at the partner FSO transceiver station.

Figure 8.13 shows how laser beams are used terrestrially as well as space-to-space, space-to-ground, and ground-to-space to transmit data. Optical links are inherently more secure and free of "eavesdropping" than typical radio or microwave communications because they are point-to-point over a tight beam. Since they are optical beams clouds, smoke, rain, snow, and other optical path obscurants do cause a reduction or loss in data transmission.

### 8.2.4 Laser Listening

When people speak inside a building, the acoustic wave generated causes the windows and other objects in the room to oscillate with the same waveform as the spoken sounds. A laser beam can be bounced off of these objects

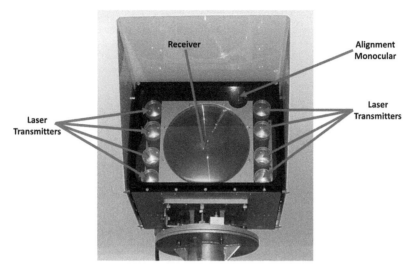

**FIGURE 8.12**
A typical laser communications transceiver.

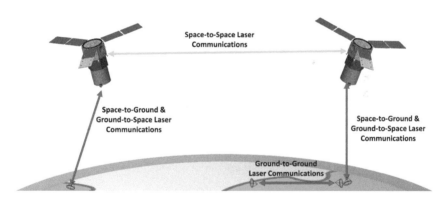

**FIGURE 8.13**
FSO uses laser beams to transmit data globally across the surface and through space.

in order to encode the beam with the sound vibrations. Those vibrations can then be recovered from the reflected beam as shown in Figure 8.14.

As the laser propagates from the aperture to the target object, the beam expands due to the divergence of the beam. The beam then bounces off the reflective object and travels back to the receiver telescope all the while it continues to spread out due to divergence. As the window vibrates back and forth the distance between it and the detector telescope is increased or decreased slightly, which in turn, increases or decreases the spot diameter on the receiver. As the spot diameter changes so does the irradiance within

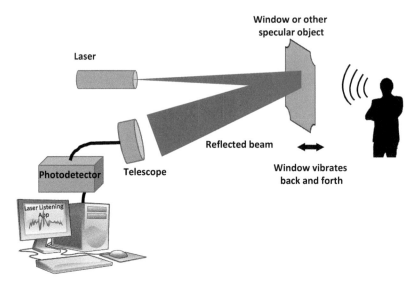

**FIGURE 8.14**
A typical laser listening device configuration.

the spot. The detector measures this as an increase or decrease in signal amplitude, which reproduces the audio signal.

### Example 8.1: Laser Listener

Consider a laser listening device with an output power, $P$, of 1 W at the aperture and a divergence of 5 mrad. The receiving detector has an aperture diameter of 2.54 cm and is located 50 m from a target window. How much will the power at the detector, $P_{det}$, vary if the maximum window displacement from acoustic vibrations is 1 mm?

The irradiance of the reflected laser spot on the detector is found by

$$I(z) = \frac{P}{\pi z^2 \tan^2\left(\dfrac{\Theta}{2}\right)}. \tag{8.3}$$

The distance, $z$, is 50 m out, plus 50 m back, plus the varying fluctuations of 1 mm due to the window vibrations. The power on the detector will be the irradiance multiplied by the detector area or

$$P_{det}(z) = \frac{P}{\pi z^2 \tan^2\left(\dfrac{\Theta}{2}\right)} \pi (0.0127\,\text{m})^2. \tag{8.4}$$

Figure 8.15 shows a graph of the power on the detector calculated by Equation 8.4. Within the 1-mm varying range due to the vibrating window the power signal varies by about 0.2 μW.

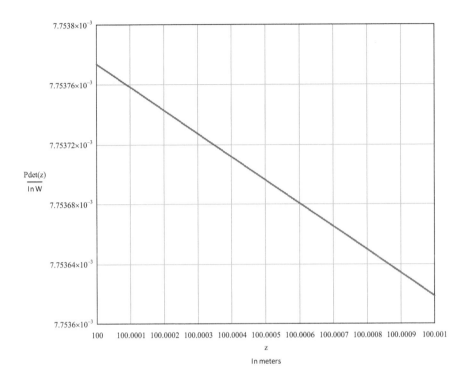

**FIGURE 8.15**
Power on the 2.54 cm diameter laser listening device detector between 100 and 100.001 m ranges.

## 8.3 The HELs

### 8.3.1 Laser Cutting, Drilling, and Welding

High-energy lasers are sometimes called HELs and offer many applications from industrial, to research, and to military. The industrial uses are probably most common and can be found around the world. HELs are often used to perform precise hole-drilling, material cutting, and even precision welding.

Typically, modern-day HEL-based cutting, drilling, or welding machines are either $CO_2$ lasers or Nd:YAG lasers. It has become a more recent trend since the early 2000s that Ytterbium, Erbium, and Thulium fiber lasers are used for industrial purposes. Machining laser systems can range from hundreds of watts to hundreds of kilowatts depending on the application.

Figure 8.16 shows a typical laser machining approach. A HEL beam is focused onto the material where it erodes away material in one location for drilling and along a path for cutting. The point where the material melts is known as the *erosion front.*

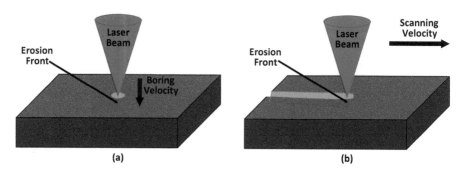

**FIGURE 8.16**
Laser machining: (a) drilling and (b) cutting.

The boring velocity or drilling rate, $v_{\text{drill}}$ for an arbitrary material is given by

$$v_{\text{drill}} = \frac{P_{\text{laser}}}{\rho A_{\text{spot}}\left(CT_v + L_v\right)}, \tag{8.5}$$

where $P_{\text{laser}}$ is the power of the laser output, $\rho$ is the density of the material, $A_{\text{spot}}$ is the area of the focused laser beam spot, $C$ is the specific heat capacity, $T_v$ is the boiling point in degrees Kelvin, and $L_v$ is the latent heat of vaporization.

For cutting the material, the laser scans horizontally across it with a scanning velocity, $v_{\text{cut}}$. The time required to make the cut to a desired depth, $z$, is

$$t_{\text{depth}} = \frac{z}{v_{\text{drill}}}. \tag{8.6}$$

The cutting velocity is

$$v_{\text{cut}} \geq \frac{d_{\text{spot}}}{t_{\text{depth}}} = \frac{d_{\text{spot}} v_{\text{drill}}}{z t_{\text{depth}}} = \frac{d_{\text{spot}}}{z t_{\text{depth}}} \frac{P_{\text{laser}}}{\rho A_{\text{spot}}\left(CT_v + L_v\right)}. \tag{8.7}$$

where $d_{\text{spot}}$ is the diameter of the beam spot size. Simplifying Equation 8.7 results in

$$v_{\text{cut}} \geq \frac{4}{z t_{\text{depth}}} \frac{P_{\text{laser}}}{\rho \pi d_{\text{spot}}\left(CT_v + L_v\right)}. \tag{8.8}$$

We should note here that both Equations 8.5 and 8.8 are simplifications and do not take into account reflection, scattering, and absorption losses in the material as it melts.

## 8.3.2 Advanced Physics Research

There are many applications in scientific research where extreme photon densities or extremely high irradiances are needed in order to study particular aspects of fundamental physics. The Department of Energy has studied HEL systems for many decades with hopes of using them to ignite fusion in reactors bombarded by multiple very HEL beams at the National Ignition Facility or NIF at Lawrence Livermore National Laboratory in Livermore, California.

Figure 8.17 shows a picture of the inside one of the world's largest and highest-energy lasers at the NIF. The system generates 192 laser beams that are focused into one tiny spot for maximum energy density. The multiple solid-state laser system produces 2.15 MJ of output energy. There are other organizations around the globe working toward higher output energy and power from laser systems.

The most powerful laser on record as of 2019 was produced by the Shanghai Superintense Ultrafast Laser Facility in Shanghai, China with an output of 5.3 PW! The Chinese laser team is designing a laser system that may push the 100 PW level with hopes that a physics concept called *pair production* can be achieved. It is possible that with high enough power density the actual vacuum of space can be "broken" and an electron-positron pair can be created. It is unclear if 100 PW lasers would be powerful enough for the experiment but sooner or later a powerful enough laser will be achieved—perhaps by one of the laser scientists or engineers studying this very text some day!

There are many other uses for HELs within the advanced physics research community. Very high-output titanium-sapphire lasers are used to accelerate plasmas in particle acceleration experiments. Some of these lasers can even be terawatt class on a tabletop reducing the cost of some particle accelerator experiments.

**FIGURE 8.17**
One of the world's largest and most powerful laser at NIF. (Courtesy DOE LLNL.)

At the University of Rochester in New York, scientists have been developing laser to act as a source of highly focused light pressure on matter to create exotic materials that normally could not be created. This area of study is called *laser energetics* and might lead scientists to answers and insights about how some planets and stars are formed.

### 8.3.3 Directed-Energy Weapons

One of the more obvious uses for HEL systems are in military applications. Since the 1970s, directed energy research organizations have sought a beamed energy weapon that can perform the way the rayguns, phasers, blasters, and DeathStars of science fiction do.

*Directed-energy weapons* or DEWs are considered ideal systems (if ever achieved) as they might have a nearly infinite magazine, engage targets at the speed of light rather than projectile velocities, and they are very discrete as some beams could be in the infrared or ultraviolet and would be invisible to onlookers.

DEW research dramatically picked up during the 1980s–1990s under the *Strategic Defense Initiative* or SDI. The goal of SDI was to develop DEWs that could engage and destroy or render useless incoming enemy *intercontinental ballistic missiles* (ICBMs) and other nuclear weapon systems. Experiments to produce large nuclear detonation-powered X-ray lasers, handheld laser weapons, portable/mobile tunable laser weapons, anti-mine and unexploded munitions laser systems, and even very high-powered chemical lasers designed to engage targets ranging from missiles to artillery rounds have been conducted over the past few decades. Modern-day research has led to solid-state and fiber HEL systems for similar military applications.

The first very large DEW was developed by the U.S. Navy and Army in the 1980s. The mid-infrared advanced chemical laser or MIRACL was a megawatt class DEW. The laser was a deuterium fluoride chemical laser and was very technically complex. Figure 8.18 shows the beam director for the laser that consisted of the major part of a very large building. MIRACL was successful in tests against various targets and experiments leading to new knowledge about how lasers could be used against ballistic missiles and anti-satellite weapons.

Figures 8.19 and 8.20 show the beam director and the general concept for the U.S. Army's tactical high-energy laser or THEL tested in the mid-1990s to the early 2000s. The laser system was a chemical laser using either hydrogen fluoride or deuterium fluoride as the gain medium. The laser was tested and shot down multiple artillery shells, rockets, and motors at the High Energy Laser System Test Facility in White Sands, New Mexico.

Figure 8.21 shows the U.S. Army's latest HEL experiment, the *high-energy laser mobile test truck* or HELMTT. The laser system to be used in the mobile DEW experiment is a solid-state fiber laser system. The HELMTT goal is to demonstrate the utility of a mobile DEW for ground forces implementation in defense against rockets, motors, artillery, and small unmanned flying systems.

**FIGURE 8.18**
The beam director for MIRACL. (Courtesy U.S. Army.)

**FIGURE 8.19**
The beam director for the THEL. (Courtesy U.S. Army.)

**FIGURE 8.20**
The U.S. Army's THEL concept. (Courtesy U.S. Army.)

**FIGURE 8.21**
The U.S. Army's HELMTT is a modern experiment to field DEWs in a mobile platform. (Courtesy U.S. Army.)

Figure 8.22 shows the U.S. Air Force's Airborne Laser or ABL. The aircraft is a Boeing 747-400F modified internally to fit all of the laser components and systems. The laser was a chemical laser system and was abandoned after 2014. A new unmanned concept is being developed with an electric laser concept.

Figure 8.23 shows the XN-1 Laser Weapon System or XN-1 LaWS developed by the U.S. Navy. The system was installed and tested on the *USS Ponce* for testing in 2014 against various threat weapons. It is a solid-state laser system that can be used to protect naval vessels against small boats, unmanned aerial vehicles, and other similar type threats.

The number of DEW implementation concepts for lasers is likely uncountable and only a few of the concepts have been discussed here. There have been many texts written with DEWs as the topic so it would be beyond the

**FIGURE 8.22**
The U.S. Air Force's ABL successfully destroyed two test missiles in 2010. (Courtesy U.S. Air Force.)

**FIGURE 8.23**
The U.S. Navy's XN-1 LaWS uses a solid-state laser to engage low-end threats. (Courtesy U.S. Navy.)

scope of this text to be exhaustive of them here. However, it is useful to have looked a few DEW systems to give the future laser scientists and engineers some perspective on this particular application.

---

## 8.4 Some Other Applications

### 8.4.1 Medical Uses

From the beginning of the field of laser science, many scientists have pondered the medical application of the laser. As it has turned out over the years, lasers have many applications when it comes to the field of medicine and human health. Lasers are used for eye surgery to correct eye lens curvature.

Lasers are used to remove tumors and to perform other types of surgeries as well as to stimulate photosensitive drugs in laser photodynamic cancer therapies. Lasers are used for many dermatological purposes such as removing skin cancers, blemishes, tattoos, and even for skin rejuvenation. Recently, lasers have been approved for hair loss treatments as well as for hair removal. Dentists also use them for scalpels in oral surgery as well as for tooth whitening. As lasers become smaller and more efficient and available in practically every wavelength, new possible medical applications will continue to be discovered.

### 8.4.2 Entertainment

Lasers are often seen at concerts and various other entertainment venues. In fact, entire industries have been developed around building and implementing lasers across multiple power regimes and wavelengths in order to create fascinating and entertaining light shows. These industries require many laser engineers to develop laser systems that can be used near people safely and generate interesting patterns and light phenomena for viewing (see Figure 8.24).

### 8.4.3 Printing

Almost everyone has come across a laser printer at some point in their lives but they likely have no idea how to print high-quality text and graphics using a laser. Figure 8.25 shows a graphic representation of the internal components of a laser printer. These printers use diode lasers to scan across a photoconductive selenium-coated drum where charges are generated in precise patterns. This charged pattern is the charged image, which then picks up toner material which is transferred to the paper.

**FIGURE 8.24**
Laser light shows are complex and require significant engineering expertise. (Courtesy Creative Commons Attribution-Share Alike 3.0 Unported license by Edward Betts.)

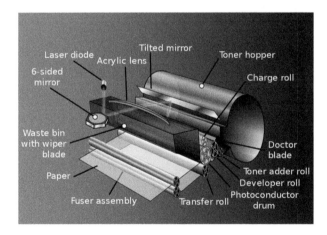

**FIGURE 8.25**
The inner workings of a laser printer. (Courtesy Creative Commons Attribution-Share Alike 4.0 International license by KDS4444.)

### 8.4.4 Too Many Applications to Count

Lasers are so versatile and come in so many shapes, sizes, output powers, wavelengths, and beam format that there are literally so many applications for them as to be uncountable. We use them every day with DVD players and burners, in grocery store bar scanners, fingerprint detection/recognition, speed detection guns, and on and on. They can be used for diamond cutting and jewelry etching. They can even be used as advanced propulsion concepts and power beaming in space. Lasers will continue to be developed and engineered for new applications and only our imaginations will limit us in what can be done with them.

## 8.5 Chapter Summary

In this chapter, we discussed many of the uses of lasers that might be implemented by the laser scientists or engineers as they begin to conduct research in the field. In Section 8.1, we discussed how to use lasers for alignment of other components and laser systems and how they can be used to probe unknown material properties and to perform precise measurements.

In Section 8.2, we discussed various applications for lasers in the field. These ranged from lidar to laser communications for leveling and surveying. It is likely that the laser scientists or engineers will need to understand details about these concepts at some point during their career.

In Section 8.3, we discussed many uses of HELs. These applications include such uses as laser machining, advanced research, and laser weapons concepts. This section gives the reader a brief overview of a very vast field of study where there are many specialization areas just as discussed in Section 8.4. There are so many types of lasers with a vast range of characteristics that there are likely an infinite number of applications for them.

Chapter 8 gives us a general overview that is a bit like standing on the beach and looking at the ocean. We realize while looking outward that the ocean is gigantic, but it isn't until one studies the ocean his or her entire life do they realize they could never learn all there is to know about it. This chapter is much like that view from the beach. There are so many subtopics and applications and concepts for lasers that we could study them forever and not learn all there is to learn about them.

## 8.6 Questions and Problems

1. Draw a standard approach for aligning a laser with a typical Fabry–Perot-type cavity with two flat mirrors.
2. How do we know that a mirror has been aligned parallel with the alignment laser's output coupler mirror?
3. Draw a Michelson interferometer.
4. What does LIDAR stand for?

5. If a mirror were placed on the Moon at 400,000 km away from earth and a green laser at 532 nm were bounced off the mirror and then back to a photodetector in a lab at earth, discuss the difficulty of measuring the vibrations on the mirror with the method used in a laser listening device. Assume that micrometeorite impacts cause the mirror to oscillate a depth of 1 μm.

6. Are there other possible techniques that might be used to measure the vibrations on the mirror on the Moon in Problem 5? Note: this is more of a discussion question and has no specific answer.

7. What if a large impactor hit the Moon imparting vibrations on the mirror from Problems 5 and 6 of more than a millimeter?

8. What is the *erosion front*?

9. Aluminum has a specific heat capacity, $C = 903$ J/kg K, $L_v = 10.90 \times 10^6$ J/kg, a density of 2,710 kg/m³, and a boiling point of 932 K. Assuming a laser drilling machine delivers 10 kW of power at 532 nm into a focal spot 1 mm in radius what is the hole drilling rate that can be achieved?

10. Assume in Problem 9 that we desire a cutting depth of 5 mm, what is the cutting rate minimum speed?

# Suggested Reading for Laser Scientists and Engineers

There are many books available about lasers, laser physics, and laser engineering, but there are few of them that are truly at the introductory level. Most of the books available are written in reference style and prove difficult to use as a first exposure to the topic. The following is a list of books (in no particular order) that may prove useful and should be in the reach of the reader once they have completed reading and working through this book. The list is by no means exhaustive but is a good point to continue following this book. In addition, as stated in the Preface there is a significant amount of information about lasers that simply cannot be found in textbooks. The laboratory, school of hard knocks, and mentorship are sometimes the best place to learn. Start with your favorite search engine on Internet and you never know where that might lead you. The trick to becoming a good laser scientist and engineer is to be able to compile all of this information as needed and to be able to understand and implement it in a useful and practical manner. Good luck.

## Books

A.E. Siegman, 1986, *Lasers*, Revised Edition, University of Science Books, Sasalito, CA.

P.W. Milonni, J.H. Eberly, 2010, *Laser Physics*, Wiley & Sons, Hoboken, NJ.

O. Svelto, 2010, *Principles of Lasers*, 5th Edition, Springer, New York.

M. Eichhorn, 2014, *Laser Physics: From Principles to Practical Work in the Lab (Graduate Texts in Physics)*, Springer, New York.

F. J. Duarte, 2015, *Tunable Laser Optics*, 2nd Edition, CRC Press, New York.

# Index

266

*Index*

Scraper mirror, 180, 181
SDI, *see* Strategic Defense Initiative (SDI)
Second harmonic generation crystal,
    152–154
Semiconductor
    diode, 162–163
    vertical-cavity surface-emitting,
        163–164
"Seven rays of the sun," 3
Shanghai Superintense Ultrafast Laser
        Facility, 247
Shells, 102
Shot noise, 84, 86–89
Siege of Syracuse, 5, 6, 10
Signal-to-noise ratio (SNR), 83, 85, 89
Simple index card, 229
Simultaneous equations, 133
Singlet states, 105–106
Skin injuries, 219–221
Slope efficiency, 203–204
Socrates, 4
Solid-state dye lasers, 160–161
Solid-state laser
    diode-pumped, 154
    fiber, 154–155
    Neodymium-Doped Yttrium
        Aluminum Garnet, 151–154
    ruby, 149–151
SOP, *see* Standard operating procedure
        (SOP)
Spatial hole burning, 130
Spatial profile, 187, 188
Spectral
    energy density, 108
    hole burning, 131
    irradiance, 83
    linewidth
        description of, 189–190
        measurement of, 190–193
        tunability, 193–195
Specular
    measurement technique
        for spatial profile, 187, 188
        for temporal profile, 180, 181
    reflections, 222–228
Speed of light, 14, 15, 19–21, 39, 40, 44, 237
Spherical laser cavity, 140
Spherical resonator, 140
Spontaneous emission, 110–112

Stability of cavity, 141–144
Standard operating procedure (SOP),
    216–217
State decay rate, 113–115
Stationary orbits, 102
Stimulated emission of radiation
    Bohr model of atom
        quantum leap, 102–105
        singlet, doublet, and triplet states,
            105–107
    Einstein coefficients, 107–110
    excited state decay rate, 113–115
    four-level model, 116–118
    three-level model, 115–116
    two-level model, 115–116
Strategic Defense Initiative (SDI), 248
Stray
    laser beams, 222
    light, 229
    reflections, 224, 227–228

**T**

Tactical high-energy laser (THEL),
    248–250
Taylor series expansion, 135
Temporal profile, 180, 181, 185, 186
TEMs, *see* Transverse electromagnetic
        modes (TEMs)
THEL, *see* Tactical high-energy laser
        (THEL)
Thermal management subsystem, 202
Thermal noise, 83, 84
Three-level model, 113–116
Threshold limit, 216
Time-dependent Schrodinger equation,
    57
Time-independent Schrodinger
    equation, 57–58, 61
Titanium-sapphire lasers, 247
Toxic hazards, 222
Transfer function, 94, 95
Transverse electromagnetic modes
    (TEMs), 188–189
Transversely excited atmospheric (TEA)
    laser, 166
Transverse modes, 187–188
*Treatise on Light* (Huygens), 19, 20
Triplet states, 106–107, 114–115

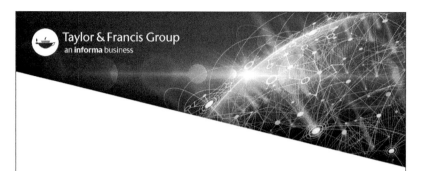